89 Advances in Polymer Science

Polymer Characterization/ Polymer Solutions

With Contributions by
M. Andreis, H. Gräger, J. L. Koenig,
M. Kötter, W.-M. Kulicke

With 89 Figures and 10 Tables

Springer-Verlag Berlin Heidelberg GmbH

ISBN 978-3-662-15083-2 ISBN 978-3-540-46012-1 (eBook)
DOI 10.1007/978-3-540-46012-1

Library of Congress Catalog Card Number 61-642

Originally published by Springer-Verlag Berlin Heidelberg New York in 1989.
Softcover reprint of the hardcover 1st edition 1989

2152/3020-543210 — Printed on acid-free paper

Editors

Table of Contents

Drag Reduction Phenomenon with Special Emphasis on Homogeneous Polymer Solutions

W.-M. Kulicke, M. Kötter
Institut für Technische und Makromolekulare Chemie
Universität Hamburg, Bundesstr. 45, 2000 Hamburg 13/FRG

H. Gräger
Horstmann-Steinberg Postfach 320, 3100 Celle/FRG

Dedicated to Professor H. J. Cantow
on the occassion of this 65th birthday

A drastic reduction of drag in the turbulent flow of solutions in comparison to the pure solvent can be observed, even when only minute amounts of suitable additives are added. This report shows that a wide range of technical and biochemical applications exists but that these applications have so far only been realized in a few exceptional cases. The reason for this must surely lie in the fact that a precise explanation for the effectiveness of drag reducing agents is neither possible from mathematical theories nor from molecular modelling.

First of all a brief outline will be given of the currently well-known theories concerning this phenomenon; molecular theories will be emphasized.

Special attention will be paid to the polymeric additives in homogeneous solutions as they can be counted amongst the most effective flow enhancers. In this respect molecular parameters (e.g., molecular weight, molecular weight distribution, solvent quality, chemical nature of the polymer, coil volume) having an influence on drag reduction will be discussed. Here the water-soluble, non-ionic polymers and polyelectrolytes are especially noteworthy because of their increasing technological and pharmaceutical importance. As a result of this work into establishing the properties required of a good drag reducing agent in homogeneous solutions, one should ask for a high degree of polymerization and a high flexibility of the chain, avoid branched structures in preference to linear ones, reduce the molecular weight of the monomer unit, and increase the coil volume, for example, by introducing ionic side groups, to name but a few examples. In addition, it has been proved that single polymer coils are effective ($c \ll c^*$). Problems arising in the characterization and handling of water-soluble substances will also be discussed.

Drag reduction decreases with flow time — which is in most application undesirable — and is obviously caused by a degradation of the polymer chain. Degradation of polymeric additives in turbulent flow cannot be easily understood on the basis of present knowledge, i.e., predictions towards the onset of chain scission cannot yet be made. These difficulties can be attributed, on the one hand, to the complex fluid structure and, on the other hand, to the fact that both shear and tensile stresses act simultaneously in turbulent flows.

Advances in Polymer Science 89

Abbreviations

Latin Symbols

a	Mark-Houwink exponent
A_2	second virial coefficient
AAm	acrylamide
c	concentration
COP	poly(acrylamide-co-acrylate)
d	diameter
De	Deborah number
DR	drag reduction
f	friction factor (Darcy-Weisbach)
f′	friction factor (Fanning)
k	Boltzmann constant

K_H	Huggins constant
k_η	constant of the Mark-Houwink relationship
K_θ	optical constant
l	length
LALLS	low-angle-laser light scattering
l_{c-c}	length of one monomer bond
LDA	laser-doppler anemometry
L_K	eddy size
L_{max}	chain length of the extended polymer
l_{mon}	length of one monomer unit
L_p	chain length
LS	light scattering
M	molecular weight
m_0	weight of one monomer unit
M_w/M_n	molecular weight distribution
n	constant of mixture distance
Δp	pressure loss due to friction
P	degree of polymerization
PAAm	poly(acrylamide)
PS	polystyrene
Re	Reynolds number
R_G	radius of gyration
$\langle R_g^2 \rangle$	mean square radius of gyration
R_H	effective hydrodynamic radius
RI	refractive index
R_θ	Rayleigh ratio
Q	flow rate
SEC	size-exclusion chromatography
t	time
T	temperature
t_k	eddy life time
t_p	relaxation time
u_m	mean velocity in pipe
u^+	dimensionless velocity
v	volume
y^+	dimensionless wall coordinate

Greek Symbols

$\dot{\gamma}$	shear rate
δ	bond angle
ε	rate of elongation
η	shear viscosity
$[\eta]$	intrinsic viscosity
η_0	zero-shear viscosity
η_{sp}	specific viscosity
λ_0	wave length

ϱ	liquid density
τ	relaxation time
τ_w	wall shear stress
ν	kinematic viscosity
χ	mixing-way constant

Indices

a	additive
cr	critical
deg	degradation
equ	equivalentσ
LS	light scattering
max	maximum
n	number average
p	polymer
s	solvent
t	time
w	weight average
θ	angle
η	viscosity average
*	critical
+	dimensionless

1 Introduction

In the area of hydrodynamics the term "drag reduction" has become a familiar name for characterizing the reduction of friction in turbulent flow through pipes. This is not caused by an improvement in the wall properties as this would only lead to a decrease in friction of a few percent. It is possible, however, to reduce the degree of friction in turbulent flow by a considerable extent, if small amounts of a suitable additive in a concentration range of only a few parts per million by weight are used; a reduction in friction of 80 percent can be reached. A tremendous change in the flow field will result, while turbulent flow still remains.

Macromolecular substances such as synthetics (Sellin 1982; Berman 1978), biopolymers (Hoyt 1985), or surfactants (Shenoy) have proved to be effective "flow improvers". Suspended fibers and solid particles (e.g., Metzner 1977, 1976; McComb 1981), however, may also produce this effect. Small suspended particles in the air ("dust") have been found to cause drag reductions of the same order of magnitude as polymers in liquids (Rossetti).

The economic advantage of this phenomenon lies primarily in the fact that turbulent flow pressure losses in applied fields add up to a considerable energy consumption. Thus, by adding a flow improver to a fluid it is possible to reduce the actual loss of pressure, so that less power is needed for pumping at the same flow rate. Alternatively, the flow rate can be raised (Fig. 1) or the pipe length increased using the same amount of power.

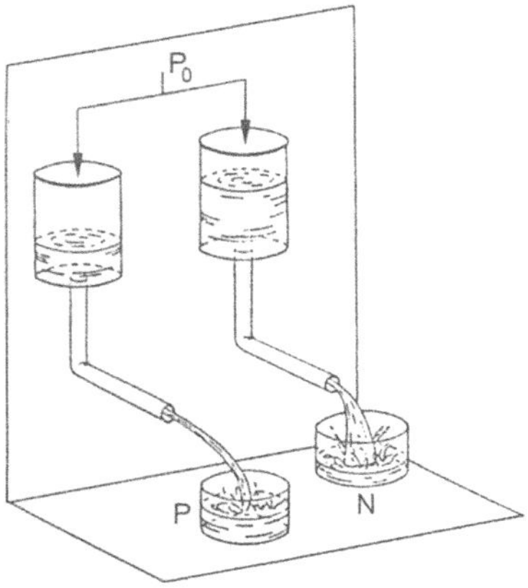

Fig. 1. Schematic representation of the different flow behavior of a NEWTONian liquid (N) in comparison to a polymer solution (P). Under equal energy conditions, it can be seen that the polymer solution in addition to a *higher volumetric flow also attains a greater length of forward travel and a higher degree of beam focusing for the jet*

The phenomenon of drag reduction is extremely interesting, not only because of its potential engineering applications but also for examining fundamental fluid dynamics, which help to elucidate the turbulence behaviour of Newtonian fluids. Specialized areas of application for this phenomenon can be found in the field of crude oil research, in pipeline flows (Burger), in fire fighting (Fabula 1971; Thorne), in increasing the speed of boats and submarines (Hoyt 1972), in water supply and irrigation systems (Sellin 1982), in the hydraulic transportation of solid particle suspensions (Golda, Interthal 1984), and in cooling and heating circulation systems (Sellin 1982). Another possible area is in the treatment of blood circulation diseases (Greene, Hoyt 1971).

The fields of application listed below (Tab. 1) are only a few examples to indicate the potential for employing the drag reduction effect.

Table 1. Areas for the technical application of polymeric flow improvers

Field of application	Advantages
Transport of crude oil	Savings in energy and investment costs (pump and construction costs)
Sewage and irrigation systems	Protection against overloading by dosage of flow accelerators, reduction of the danger of cavitation, saving in pumping energy
Fire fighting, as well as hydrostatic cutting tools used in the textile, food, and mining industries	Higher efficiency due to increased jet velocity and beam focusing
Shipping industries	Increased velocity and fuel saving
Aircraft refuelling	Reduction of fuel atomisation and risk of explosion
Long range heat transport systems	Reduction of heat loss and saving of pumping costs
Medicine	Addition to the blood circulation system during treatment of arteriosclerosis
Solid fuel transport	Increased throughput in pipelines

"Although the mechanism producing drag reduction is not yet known, a number of polymer characteristics making for good drag reducers have been determined. A long-chain backbone and flexibility are important characteristics of good drag-reducing agents. For instance, of two polymers with the same molecular weight and same structural units, a linear one will be more effective than a highly branched one. Also, for two different polymers of similar configuration and the same molecular weight, the one with the lower-molecular-weight monomer will have the greater drag-reducing effect if both are utilized at the same weight concentration" (Bird). This citation reflects the present state of knowledge. As so far only very little is known about the influence of molecular parameters on the effectiveness of drag reduction and the stability of macromolecules in turbulent flow.

In this paper particulary strong consideration will be given to investigations recently made in this field by our working group. Important parameters include the average molecular weight, molecular weight distribution, charge-density distribution, copolymer composition, chemical nature of the additive, thermodynamic quality of the solvent, etc. Therefore particularly strong emphasis will be given to these aspects in this article.

This report is based essentially on earlier review articles by Shenoy (1984); Sellin, Hoyt, Scrivener (1982); Giesekus, Bewersdorff, Dembeck, Kwade, Martischius, Scharf (1981, 1985); and furthermore, Berman (1978); Dschagorowa, Mennig (1976); Virk (1975); Little, Hansen, Hunston, Kim, Patterson, Ting (1975); Hoyt (1972, 1974, 1986); Lumley (1973); and Patterson, Zakin, Rodriguez (1969). This list of authors is by no means complete considering the numerous publications already issued on this topic.

1.1 Historical Background

The first reports on the drag reduction phenomenon are found in publications by Blatch (1906), Forrest (1931), as well as by Brautlecht (1933), who measured the flow behaviour of paper pulps. In independent studies made by Mysels (1949) and Toms (1948) this phenomenon was observed in the turbulent flow of gasoline in pipes when aluminum soaps were added and, in Toms experiments, when polymethylmethacrylate in monochlorobenzene was used. The reduction of friction is therefore often termed the "Toms effect". This flow behaviour is also known in the literature as the "Non-Newtonian-", "visco-elastic-", "Texas-" or "Texas-Toms-effect", due to the contributions made by Texan researchers. It is also more generally known, and this term will be used in this paper exclusively, as the drag-reduction effect of flow additives.

In 1959 Dodge's thesis under the supervision of Metzner was published at the same time as Shaver's findings under the supervision of Merrill. Both noticed unusually low friction factors for certain non-Newtonian solutions like those of sodium carboxymethylcellulose in water. At around the same time, industrial researchers made similar observations with certain additives, the most prominent being guar gum, which is a polysaccharide derived from a plant. The gums were used to suspend sand in the sand-water mixtures utilized in oil-well fracturing operations.

Due to the extraordinary effectiveness of polymers and the low concentrations needed, extensive investigations have been made on synthetic polymers. In 1964,

Savins demonstrated the potential of reducing energy loss and coined the term "drag reduction". Since then, a great number of studies have been carried out in order to determine the parameters influencing the degree of drag reduction, especially in the case of polymers. The drag reduction effect has been examined in organic and aqueous solutions. Polyethyleneoxide has been used primarily since it has proved to be the most effective polymer, followed by polyacrylamide and its partially hydrolyzed derivatives, as well as the natural substance "guar gum". This effect surprisingly also occurs in nature; for instance, the skin of some fishes produces substances which reduce friction and therefore allow a speedy retreat (Rosen, Lagerstedt).

2 Flow Improvers in Pipelines

It appears that the first successful large-scale application of polymers as flow improvers was in the field of crude oil research (Ram). For instance, "guar gum" was used under carefully defined conditions in the transportation of oil in pipelines with a diameter of 14 cm, thereby lowering the energy level of the pump to factor 5. Flow improvers are also used in crude oil transportation in the Trans-Alaska-pipeline (Hoyt 1972; Burger) and in pipelines from offshore platforms to shore facilities (Hoyt 1986; Beatty). Here polymers ranging in concentration from 5 to 25 ppm have been used in pipes with a diameter up to 1.20 m (Burger). The possible levels of drag reduction were up to 28%, under laboratory conditions values up to 80% have been possible.

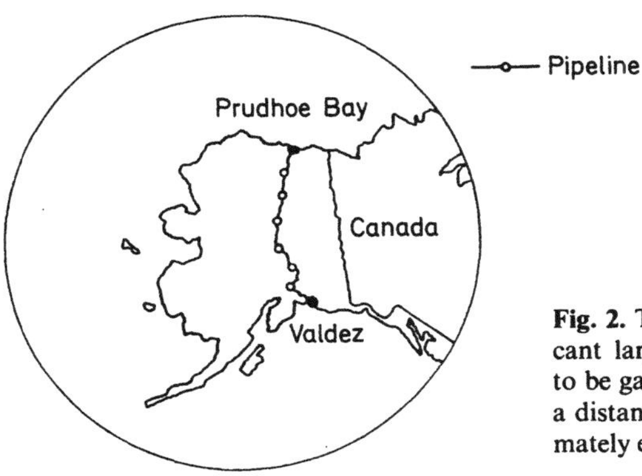

Fig. 2. The Alaska pipeline, one of the most significant large-scale technical examples of the benefits to be gained from the "drag reduction" effect. Over a distance of 1287 km, polymer is injected approximately every 100 km at a concentration of 5–25 ppm

Large-scale applications are limited because of the sensitivity of polymers to chemical, thermal, and/or mechanical degradation (Martischius). Thus, the use of polymers should either be limited to open-ended systems with short distances to be covered (such as in the use of water hoses in fire fighting) or an additional injection device must be used in pipeline transportations, to automatically adjust the content of (fresh) flow improvers in case of decreased effectiveness.

Copolymers have been employed in organic solvents and have increased the flow rate of oil by 50% (in pipes with a diameter ranging from 15–50 cm and with a concentration level up to 100 ppm) (advertising pamphlet Chemlink). It is not necessary to

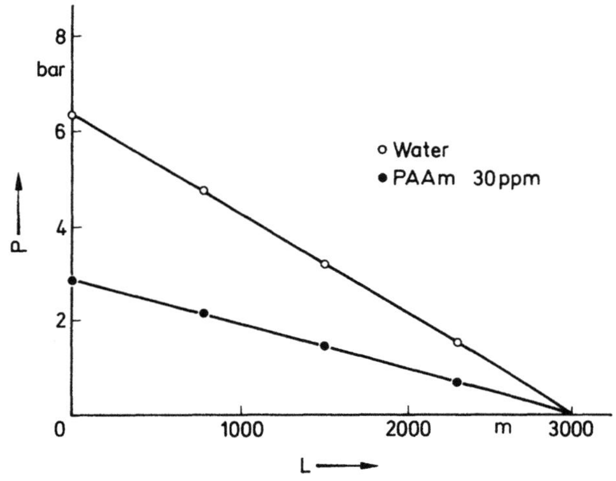

Fig. 3. Pressure drop measurements along a pipeline (diameter = 0.3 m, length = 3000 m, flow velocity 2.7 m/s). Conveyance of water and of an aqueous solution of polyacrylamide (Durst 1982; Interthal 1985)

separate the polymers afterwards since the amounts lie below the level of contamination, and do not show any other negative effects.

Injection of small amounts of high-molecular-weight polymers in the turbulent flow range can lead to substantial drag reduction in pipelines carrying low-molecular-weight liquid hydrocarbons such as kerosine, diesel fuel, crude oil, or heavy crude oil (Schmerwitz). For example, 20 ppm poly-(1,2-butyleneoxide) with a molecular weight of $2.05 \cdot 10^6$ g/mol were added to Alaskan crude petroleum, resulting in 45% drag reduction (Carville).

Other examples of this kind are reported in the literature (Durst 1982). Figure 3, for example, shows the results of pressure drop measurements taken on a pipeline having a diameter of 0.3 m and a length of 3000 m. The addition of 30 ppm polyacrylamide (PAAm) reduces the pressure drop from 6.2 to 2.9 bar.

Another possible technical application may be the addition of polyacrylamide, for example, to lower the flow resistance in diffusers (Povkh).

3 Definition of Equations Used to Describe Drag Reduction

A generally and widely accepted definition of drag reduction was provided by Savins. He defined this effect as the increase in pumpability of a fluid by adding certain materials to a fluid flowing under turbulent conditions. Drag reduction for a pipe is given by the following equation:

$$DR = \frac{(\Delta P_s - \Delta P_a)}{\Delta P_s} \quad \text{(at constant flow rate)} \tag{1}$$

DR = Drag Reduction
ΔP_s = pressure loss due to friction in unit length of the pipe for solvent alone
ΔP_a = pressure loss due to friction in unit length of the pipe for the solution containing the additive

Drag reduction can also be computed by using the following relationship:

$$DR = 1 - \frac{f_a}{f_s} \quad \text{(at constant flow rate)} \tag{2}$$

f_a = friction factor of the additive solution in unit length of the pipe
f_s = friction factor of the solvent alone in unit length of the pipe
The friction factor f (Darcy-Weisbach) is defined by:

$$f = \frac{8 \cdot \tau_w}{\varrho \cdot u_m^2} \tag{3}$$

τ_w = wall shear stress
ϱ = liquid density
u_m = mean velocity in pipe

$$f(\text{Darcy-Weisbach}) = 4 \cdot f'(\text{Fanning})$$

Other possibilities of calculating the drag reduction effect are the following:

$$DR = 1 - \frac{\tau_{w,a}}{\tau_{w,s}} \quad (u_m = \text{const}) \tag{4}$$

$$DR = 1 - \frac{u_{m,s}^2}{u_{m,a}^2} \quad (\tau_w = \text{const}) \tag{5}$$

4 Proposed Theories for the Drag Reduction Phenomenon

4.1 Wall Effects — Shear Thinning

Experimental evidence suggests that the region near the wall plays the main role in drag reduction occurring in turbulent pipe flow of very dilute solutions of macromolecules. Oldroyd indicated that the interaction between elastic macromolecules and structure elements in turbulent flow would have to play an important part. His "*slip-at-the-wall*" *theory* has remained a mere engineering correlation method, rather than a fundamental answer to the mechanism of drag reduction, and will remain so until an improved method for the measurement of velocity profiles near the wall is found. Based on Oldroyd's theories, Toms proposed the idea of a *shear-thinning wall layer* with an extremely low viscosimetry, which resulted in lower friction coefficients for the drag reducing solutions than for pure solvents. But looking at the rheograms of drag reducing polymer solutions, it was evident that they were not shear-thinning but, in fact, Newtonian by conventional viscosimetric methods. This theory was in contrast to Walshs results. He showed that solutions of polymethacrylic acid, which is essentially a shear-thickening substance, gave considerable drag reduction. Only in high concentrations of polymeric additives could non-Newtonian effects (in shear flow viscosity) be found, and then, the shear thinning led to solution viscosities much higher than that of the pure solvent. Furthermore, Lumley (1964) showed that turbulence is not sensitive to shear-induced changes in viscosity, which is in turn dominated by inertial forces.

4.2 Adsorption Effect

Another kind of "wall-effect" was proposed by El'perin (1967). He suggested that an adsorbed layer of polymer molecules could exist at the pipe wall during flow and this could lower the viscosity, create a slip, dampen turbulence pulsations, and prevent any initiation of vortices at the wall. Later work (Little 1969), however, with a transparent pipe and dyed polymer, showed that the "*adsorption*" could in be fact an experimental artifact (a quantity of polymer solution, trapped in pressure gage piping, slowly diffused back into the solvent flow). Although polymer molecules do more or less adhere to clean surfaces in thin films, there is no interaction with the bulk of the solution which could alter the flow properties (Gyr, 1974). Thus, it is evident that adsorption of the additives on surfaces is not the reason for the drag reducing effect.

4.3 Normal Stress — Non Isotropic Viscosity

The existence of a non-isotropic viscosity was thought of as another possible way of explaining this phenomenon. This so-called "*non-isotropic viscosity*" is a viscosity which is low in the direction of the flow stream lines but high enough in any other direction to dampen a substantial percentage of the oscillation leading to turbulent fluctuations. *Differences in normal stresses* could act as a mechanism, but Gadd (1966) concluded from his own measurements with polyethyleneoxide, polyacrylamide, and guar solutions that there was no obvious correlation between effectiveness in reducing turbulent drag and the differences in these normal stresses. Patterson (1968) used a Maxwell model to calculate reductions in turbulence intensities in a series of polymer fluids, compared with the solvents. This reduction was then compared with the observed drag-reducing ratios obtained with the polymer solutions. The viscoelastic model required evaluation of the normal stress difference which was obtained experimentally by measuring the thrust from a jet of the polymer fluid. The normal stress difference was significant, but the solutions employed were extremely concentrated by drag-reducing standards, and hence very obviously viscoelastic. Whether or not these calculations would be applicable to the extremely dilute concentrations utilized in drag reduction experiments seems questionable. For the solutions employed, the drag reduction results almost comply with the calculations. But, it seems that all theoretical calculations can only provide a general background in the study of friction-reducing polymer solutions. So the application of a simple Maxwell model or a distribution of Maxwell elements cannot account for the observed changes in viscous and elastic values with change in turbulence frequency.

Boggs combined, in a mathematically elegant approach, a constitutive equation, including normal stress terms with equations of motion to form differential equations similar to the Navier-Stokes equations. He found that viscoelasticity had a destabilizing effect and concluded that transition should occur earlier. Destabilization of the flow may explain the observations of "early turbulence" (Forame); the increased drag observed before normal transition may be due to secondary motions arising after an instability. They, however, suggested that in the turbulent region frictional drag was a function of one-third the power of the Weissenberg number, which is the ratio of the elastic forces to the viscous forces. Later, Lockett calculated that viscoelasticity leads to destabilization, but he inferred that a vortex component was introduced in the flow direction. Furthermore, Black developed a new theoretical model of turbulent shear flow which indicated greater sublayer stability, but showed an increase in the longitudinal velocity fluctuations when polymers are present during the flow. Ruckensein suggested that drag reduction could be due to two effects of viscoelasticity:

1) Using a Maxwell model as a constitutive equation for a viscoelastic fluid, one can show that the instantaneous shear stress at the wall is smaller in the viscoelastic fluid than in the corresponding Newtonian fluid.
2) The renewal of the elements of liquid along the wall takes place more slowly in the case of viscoelastic fluid than in the case of a Newtonian fluid.

A detailed discussion of the relationship between drag reduction and the random surface renewal in turbulent pipe flow is available in Fortuin's report.

4.4 Decreased Production of Turbulence

Another proposal to explain drag reduction is a possible relation between viscoelasticity and energy dissipation. Astarita suggested that turbulence in viscoelastic liquids is perhaps not suppressed, but is less dissipative than in viscous liquids. Gadd (1965, 1966) proposed that the cause of drag reduction was not due to reduced turbulence dissipation, but rather to *decreased production of turbulence.* The experiments of Johnson show the decreased production of small eddies in a developing boundary layer containing polymers, which gives further support to the suggestions that polymer additives interfere with the production of turbulence.

Walsh, in his very comprehensive theory of friction reduction, starts by considering the small disturbances in the viscous layer which grow by extracting energy from the local velocity profile through the action of the Reynolds stresses. The disturbances tend to lose energy due to viscous dissipation. For polymer solutions, Walsh believes these small disturbances to tend to store energy in the polymer molecules. If a small disturbance extracts more energy locally than it loses, it will grow. The disturbances, or vortices, so generated move out from the wall as they are convected downstream. Thus, small disturbances at the edge of the viscous sublayer become part of the structure of the turbulence in the outer part of the boundary layer which in turn become responsible for the Reynolds stresses of the turbulent flow. The most important aspect of this theory is that the large-scale disturbances which produce the Reynolds stresses some distance downstream were, at an earlier time, small disturbances at the edge of the viscous sublayer some distance upstream. Once the small disturbances start to grow, the effect of the polymer molecules may be neglected. All the polymer molecules do is to slightly alter the energy balance of the turbulent fluctuations close to the wall. By slightly altering this balance, the molecules allow viscous dissipation to destroy disturbances which would have had sufficient kinetic energy to grow. By decreasing the number of disturbances which grow per unit area with time and move out from the edge of the viscous sublayer, the addition of the polymer molecules ultimately changes the structure in the outer part of the boundary layer, resulting in lower Reynolds stresses and hence friction reduction. Kilian has also shown that in drag-reducing polymer flows the turbulence energy is considerably reduced at high frequencies and is essentially due to the decreased reproduction of turbulence. Walsh's theory does not predict turbulence dampening in a free flow and is based upon the assumption that the phenomenon is essentially due to the existence of a wall-boundary-layer flow. This concept was supported by the experiments of Walters, who injected polymers from a porous-walled pipe section in flowing water and observed drag reduction occurring in the downstream sections of the pipe where the polymer was in contact with the wall and pure water was flowing through the interior of the pipe.

4.5 Injection Experiments

The first injection experiment, in which polymer was injected into a pipe, was described by Wells.

He found that during the injection of polymer solution into the turbulent core, no reduction in local pressure gradient occurred until the polymer was diffused into the wall region. In later injection experiments by Vleggaar, a 0.5% solution of polyacrylamide was injected into the center of the pipe, where it formed a long thread which remained intact over a distance of more than 200 tube diameters ("heterogeneous" drag reduction). Although the most important part of the injected polymer was not dispersed, a large degree of drag reduction occurred, which was two to three times higher than of a homogeneous premixed solution.

Tullis explains this phenomenon with the existence of undissolved globules of polymers, which are several orders of magnitude larger than individual polymer molecules. In this concentration range it is highly possible that aggregates or clusters of polymer molecules are formed and their particle weights or their dimensions are much higher than for a single polymer molecule.

Goren found asymptotic behaviour, which is higher for this "*heterogeneous*" *drag reduction* than for a premixed homogeneous solution, whereas Tullis, Ramu, and Maus reported only little influence of the concentration of the injected polymer solution on this asymptotic value. Walters and McComb (1982) found no such influence.

Heat transfer measurements for this "heterogeneous" drag reduction (it should be noted, however, that in long tubes this type of drag reduction will gradually transform into the homogeneous one) demonstrated that in Vleggaar's experiments the heat transfer reduction could be correlated with drag reduction. Sellin (1974) also observed that, by injecting concentrated polymer solutions, turbulence was suppressed before the polymer had reached the near-wall region. All these experiments seem to suggest that the core-injection of concentrated polymer solutions results in a different type of drag reduction than obtained with dilute homogeneous polymer solutions.

McComb (1982) investigated both types of injection; namely at the wall and at the center line of the flow. In their apparatus the injection point was situated where the flow was well developed. These experiments led to the following conclusions: when polymer solution was injected at the center line of the pipe, the local drag reduction increased with distance downstream from the injection point. This was due to the polymer spreading out radially, as the injected solution was carried downstream. When the polymer solution was spread uniformly across the pipe, the local drag reduction attained its constant, asymptotic value.

When polymer solution was injected at the wall of the pipe, the local drag reduction again increased with distance downstream, but at a more rapid rate than when injection was at the center line. Also, the variation of drag reduction with downstream distance tended to show a slight overshoot before falling off to the same asymptotic level as in center line injection.

For both types of injection, the asymptotic drag reduction was greater than obtained with homogeneous solutions under otherwise identical conditions. This was particularly noticeable at lower Reynolds numbers. It was attributed to a reduced onset wall shear stress and McComb believed that this was caused by the presence of molecular aggregates rather than by individual molecules.

Stenberg also injected concentrated polymer solutions into pipe flow, but in his apparatus the injection was at the inlet of the pipe, via a rotating impeller mixer. Pressure drops were measured when the mixer was inoperative and also when it was running at various speeds. With no mixing or only poor mixing, the results comply with those of Vleggaar and show reduced onset stress. However, with efficient mixing, the usual behaviour of homogeneous solutions is regained. Dye visualization and schlieren photographic studies revealed the presence of small visible polymer strands which disappeared when the mixer was used. Stenberg concluded from his results that there was no essential difference between homogeneous and heterogeneous forms of drag reduction, but, the question of where the polymer molecules act in the two cases was not considered.

In his experiments with centrally injected polymer solutions, Bewersdorff (1984) showed in flow visualization experiments that the injected polymer solution formed a liquid thread which remained intact down the entire length of the pipe ($l/d = 270$), provided that the concentration of the injected polymer solution exceeded 0.3%. This polymer thread is deformed by turbulent flow, i.e., it can undergo elongation by stretching and even from loops (see Fig. 4).

This was in contrast to a glycerine solution having the same zero-shear-viscosity which was dispersed after a few pipe diameters at a relatively low Reynolds number. For injection concentrations of 0.25% and 0.3%, the polymer thread did not remain

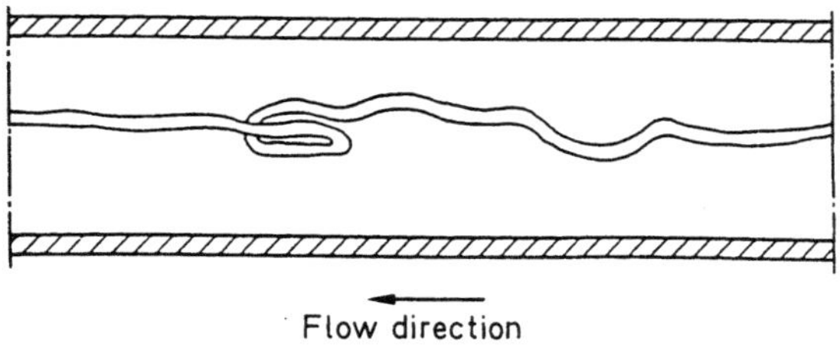

Fig. 4. Examples of the behaviour of a polymer thread (c_p = 500 ppm, Re = 70000) in turbulent pipe flow. Pictures taken on a high speed camera at a frequency of 1000 exposures per second serve to illustrate this effect (Bewersdorff 1984)

intact, and using the flow visualization method, Bewersdorff could show the existence of stable polymer strands which were distributed over the whole cross-section. Although there was no important mixing of the injected polymer solution, remarkable drag reduction was recorded, which was greater than for monogeneous solutions. For injection concentrations below 0.2%, no polymer threads or strands were observed. Drag reduction increases much more quickly with lower concentrations and the asymptotic value of drag reduction at the end of the measuring section is nearly identical with that of a homogeneous polymer solution. From his velocity profile measurements, Bewersdorff found large differences in the slope of the core region between injection experiments and homogeneous solutions! For the homogeneous ones, the velocity profiles were nearly parallel to those obtained for water, and in the case of the injection experiments the velocity profiles show a parallel shift with increasing Reynolds numbers and they are not parallel to the Newtonian line. In Bewersdorff's opinion, this is an indication of a change in the structure of turbulence in the core region, because Prandtl's mixing length constant is decreased in the injection experiments. Berman's (1986) observations imply that non-homogeneous drag reduction involves a modification of the large eddy structure. Smith investigated the influence of drag reducing agents on mixing in turbulent pipe flow and found that with increasing concentrations of polyacrylamide the mixing length increases exponentially. This dependence is greater than expected from the decreased energy dissipation resulting from the reduced drag. The effect of the Reynolds number is less significant than that of the polyacrylamide concentration.

With the aid of the two-color Laser-Doppler-Anemometry (LDA), Bewersdorff was able to measure the axial and the radial turbulence intensities simultaneously and also the Reynolds shear stresses. The injection of polymer results in a damping of both intensities in the region of their maxima. In his Reynolds shear stress measurements he showed that the polymer injection results in a drastic damping, and the stress maximum is shifted towards the center of the pipe. In a homogeneous polymer solution the maximum of the Reynolds shear stress remains in the same position as for water. Only in the region of the buffer zone are the shear stresses reduced.

In contrast to center-line or direct wall injection, it was found by circumferentially injecting a ring of relatively concentrated viscoelastic fluids into the turbulent boundary layer in pipe flow that drag reduction is considerably higher than in homogeneous solutions of the same average concentration, even at large distances from the injection point (Frings 1984, 1988; Usui). Velocity profile measurements demonstrate in the case of near-wall injection that the turbulent structure could be altered in the near-wall and also in the core region of the pipe flow, indicating that the polymer lumps and threads created by the near-wall injection influence a much wider spectrum of turbulent eddies than either center-line injection or homogeneous drag reduction, in particular (Frings 1988).

4.6 Vortex Stretching

A further possible mechanism which can be the cause of drag reduction is the development of a resistance to *vortex stretching* due to the presence of the additives. Gadd (1965) has suggested that vortex stretching action occurs along the wall, the resistance to vortex stretching is postulated to give reduced mixing and more rapid decay of eddies.

He has pointed out that the larger polymer molecules, with larger relaxation times, would probably affect the large eddies with less intense stretching rates and cause them to decay more rapidly. Further support of vortex stretching inhibition, on the basis of large elongational viscosity, has been postulated by Kuo in his grid turbulence analysis.

Gyr (1968) pointed out, on the other hand, that the process of vortex stretching inhibition can only be effective in the case of the smallest eddies near the wall. Metzner (1970) has shown the remarkable resistance to elongational strain of dilute polymer solutions, the implication being a resistance to vortex streching. Lacey indicated that there was a net macromolecular movement towards the center of the vortices, followed by a resistance to stretching due to an interaction in the long overlap regions of aligned molecules. Gordon (1972) also explained the drag reduction phenomenon by a resistance to vortex streching caused by filament formation in drag reducing polymer solutions.

4.7 Molecular Stretching — Elongational Flow

Another possible explanation for drag reduction involves the idea of *elongational flow*. It is quite likely that the addition of small quantities of polymer to a solvent leads to a substantial increase in the resistance to elongational flow, thereby resulting in less turbulent bursts and thus lowering the turbulent drag. Peterlin expanded this idea in such a manner that the statistical coiled polymer molecules were stretched strongly in areas with elongational flow character, which probably exist between adjacent eddies. Shin suggested the possibility of anisotropic viscosity effects due to coil extension of polymer molecules.

Tulin visualizes the polymer molecules becoming greatly extended in the shear direction, thus providing a stiffening effect, which absorbs energy from the turbulent eddies and radiates it away as elastic shear waves, which later decay due to viscosity. He verified from the experimental data of Hoyt (1964) that the turbulent dissipation in the polymer solution is proportional to the product of the concentration and the square root of molecular weight. Furthermore, he argued that the increased turbulent dissipation, which could be as high as three times that of a normal fluid, was proportional to the increase in the sublayer thickness of the boundary layer and thus resulted in drag reduction. Lumley (1967) concurred with Tulin's theory that the growth of the laminar sublayer could be explained by molecular extension, but he found by calculation that the elongation itself is very slight. Based on a number of experiments, Cottrell found that there was only a very slight macromolecular elongation during shear flow. Measurements in porous media flow by Kulicke, Haas; Haas; Durst (1984), and Hashemzadeh also showed that the elongational viscosity of polymers increases only very slightly. In Bird's theory, the elongational viscosity should increase in the range of three decades, if the polymer molecules could be stretched to their full extent. In porous media, deformation or elongation is only slight, because mechanical degradation occurs, and thus the increase in elongational viscosity lies in the range of one decade. Keller showed that high chain extension, observed experimentally as flow birefringence, can only be achieved in persistently extensional flows. The chain extension can be observed in the double jet and four or six roll mill apparatus.

Since the elongation in Cottrell's experiments appears to be so slight, Lumley (1967) proposes molecular entanglement as a means of interfering with, and enlarging the sublayer eddies. Entanglements extending cross-stream are visualized as resisting the formation of the streamwise vortices. But Patterson (1970) as well as Oliver found pipe drag reduction with polymer solutions as dilute as 0.03 ppm; with the molecular separation distances and volume fractions involved it is quite unlikely that molecular entanglement could play any part in drag reduction. On the other hand, Ellis showed that polymer solutions stored for several weeks exhibited a lower level of drag reduction than freshly mixed ones, thus implying that a molecular disentanglement had taken place during storage. Dunlop discussed several methods of detecting and characterizing molecular aggregates in solution and showed that several polymer solvent systems of interest in drag reduction form aggregates. In the case of polyacrylamide they demonstrated in a scanning electron microscope that a 10 ppm aqueous solution shows a complex fibrillar structure indicating a high degree of entanglements. Hence, they come to the conclusion that molecular aggregates occur in polymer solutions even at high dilutions, and that many polymers which can produce drag reduction also form clusters or aggregates. Furthermore, aggregates may often be found by shearing, that is, in concentrated solutions at least.

However, according to Little (1969), polyethyleneoxide solutions of different molecular weights gave the same drag reduction when their concentration was proportional to the critical concentration at each molecular weight (i.e., the computed concentration for the polymer coils to touch each other). Kinnier obtained similar results but used the concept of "equivalent concentrations". He found that to have equal drag reduction for different molecular weight polymer solutions, one has to have equal volumes of polymer based upon the hydrodynamic sphere considerations.

Pfenninger suggested that the action of additives is to interfere with and weaken the disturbance vorticity by stretching and breaking up the polymer molecules. The molecules partially absorb the kinetic energy of the vortices, allowing them to grow larger and flow further away from the wall before becoming unstable, thus resulting in thickened laminar sublayers and reduced friction.

A similar model by Peterlin illustrates a method of turbulence suppression by mechanical interference. One end of the macromolecule lies in the core of a "microvortex" and the other outside; this strains the intermediate section causing molecular exten-

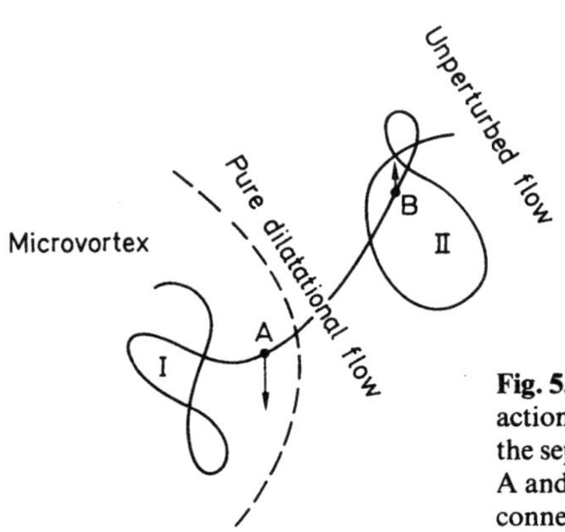

Fig. 5. Molecular stretching. A possible molecular stretching action of polymer molecules in drag reduction. Forces on the separated coil section I and II, represented by the vectors A and B, cause an extension of the macromolecular segment connecting the two regions (Peterlin)

sion, absorption of energy, and dissipation of the vortex. Thus, the polymer molecules help to stabilize the surface layer by mechanical interference, which suppresses the turbulence and reduces the growth of the vortex. Furthermore, using Corino's picture of the turbulent bursting process, which is initiated in the sublayer and mainly responsible for the turbulent energy production, Gordon (1970) reported that polymer molecules suppressed the turbulent bursts and thereby increased the resistance to stretching. He suggested that, if the ejection frequency of turbulent bursts is unchanged in polymer solutions, the resistance to stretching inhibits the magnitude of the bursts and hence the flow becomes less turbulent. Latto's measurements indicated that there is a reduction in the size range of turbulent eddies when a polymer is present in the flow. Gyr (1975) indicated a reversal in the relative occurrence of the streamwise and spanwise vortices in the presence of polymers. His results comply well with those of Brennen as well as those of Fortuna. A decrease in the velocity fluctuations in the spanwise direction was also observed by Rudd (1971, 1972). Tiederman reported that at streamwise locations along the direction of flow where drag reduction occurs and where an injected fluid is not yet uniformly mixed with the flow, the dimensionless spanwise streak spacing increases and the average bursting rate decreases.

Kim (1973) made investigations on polyacrylamide which he produced himself. By entering ionic side groups in the polymer backbone chain, so causing an extension of the molecules due to electrostatic charge repulsion, they found drag reduction effectiveness proportional to the ionic content of these side groups in the polymer molecules. Later, Banijamali confirmed his results with investigations into polyacrylic acid which he produced himself. In the case of ionic additives especially, an important loss in drag reduction during flow can occur, which is due to these additives being adsorped onto solid surfaces. This includes the pipe wall as well as any suspended particles in the flow. The extent to which this occurs will, for instance, depend on the surface chemistry of the system and in particular also on the ionic character of the polymers used. Anionic or, more commonly, cationic polymers will of course flocculate suspended particles of the opposite ionic character, but they will also bond with dissolved metal ions, particularly bivalent ones such as magnesium and calcium. In this way, all the dissolved polymer molecules may be precipitated from the solution, if the pH is correct, and can so suppress the drag reduction effect completely. Providing that nonionic polymers are used, the presence of metal ions in solution, and to a lesser extent suspended fine particles, is unlikely to have a serious effect on engineering applications for friction reduction. However, laboratory tests may be needed before selecting polymers for use in heavily polluted flows or before considering using strongly ionic polymers in other liquids.

By referring to the above results, a fundamental relationship between the macromolecular and the turbulence scales was recognized as important at an earlier stage, despite the fact that in dilute solutions the conventional viscoelastic properties of these solutions are not apparent. Several authors have tried to relate drag reduction to the characteristic relaxation time, but with only limited success. Better results were obtained by Virk (1967) who demonstrated that onset occurs at critical wall shear stress and that this is related to the dimensions of the polymer molecules normally defined by the radius of gyration. This suggested that the onset of drag reduction depends only on a length scale parameter, but the majority of experimental data does not show any clear correlation between the length scale of the polymer molecule and its characteristic turbulent flow (Lumley 1969). Virk (1971) had, however, constructed a *three layer model* which conforms with the maximum drag reduction asymptote in the friction factor diagram. The three layers in the velocity profile are the usual viscous sublayer, an elastic intermediate layer (the buffer layer) where the mixing length constant is derived from the maximum drag reduction asymptote, and an outer (core) region with a Newtonian mixing length constant. In this three layer model, the turbulent Newtonian core region diminishes if the reduction in friction increases. The maximum drag reduction effect is achieved when the buffer layer extends over the whole pipe cross-section, with the exception of the viscous sublayer. This limiting feature is described as the "maximum drag reduction asymptote" or the so-called "Virk's asymptote" which has been confirmed by numerous experiments with polymers as flow improvers.

In the case of the drag reduction mechanism, it seems more probably that a relationship between the molecular time scale and the turbulent flow characteristic time exists. Fabula (1966) was the first to give an onset criterion based on the interaction between flow time scale and molecular relaxation time. This time scale hypothesis was confirmed by Berman (1977), showing that both onset and drag reduction curves are related to the relaxation time. More recently, several authors have been able to demonstrate the importance of the elasticity of the macromolecules in explaining drag reduction by polymer additives. Both Lumley (1977) and Berman (1978) have shown that these molecules may become extended in turbulent flow if regions of predominantly extensional flow (as opposed to rotational motion) exist for long time spans. Otherwise, the randomly coiled polymer molecules may be rotated by the flow to a fresh orientation before the elongational strain rate has persisted long enough to extend them completely. This explains the absence of drag reduction in laminar flows as this is fully rotational in character with no regions of elongational strain. In turbulent flow, the molecular stretching is initiated by the magnitude of the non-rotational strain rate, while the extent to which the molecule is stretched depends upon the persistence of the elongational strain. These conditions are found to exist in the small region during the turbulent bursting process, as has been demonstrated experimentally by Falco. The existence of these regions of abnormally high elongational viscosity can be used to explain a number of observed effects produced by drag reducing macromolecules, these are:

i) a thickening of the elastic sublayer
ii) changes in the turbulence power density distribution
iii) changes in the level of longitudinal turbulence components.

Walsh's theory considers a drag reduction taking into account the energy stored in the stretched macromolecules. He proposed that the small-scale disturbances near the wall tend to grow, because they extract energy locally from the flow much faster than they can dissipate it. Polymer molecules, by storing this excess energy, can prevent the growth of these small-scale instabilities. The experimental results of Berner are consistent with this model in that they show a decrease in the turbulent dissipative energy near the wall and an increase in the Taylor microscale. Kohn also devised a model which took the energy stored by the macromolecules into consideration, but neither of these models succeeded in overcoming the difficulty of placing a value on this stored energy. Furthermore, the problem of finding a mechanism for recycling stored energy to the flow remains unsolved.

Durst (1982) proposed a basic mechanism leading to a general explanation, not only for the increase of pressure drop in porous media flow, but also for the decrease in turbulent wall flow (drag reduction). These two effects are expected to be related to a marked stretching of the dissolved high-molecular-weight polymers in elongational flow is theoretically explained on the basis of the Fene model for (finitely-extendable, non-linear-elastic) dumb-bells (Bird; Warner). In porous media, the elongational flow character, caused by converging and diverging flow passages, can be described by the ratio of the superficial velocity to the bead diameter of the porous media. The elongational flow fields occurring in turbulence, e.g., produced between counter-rotating pairs of eddies, are formulated in a similar way by the ratio of the root-mean square cross velocity to the length of the most energetic eddy. In order to simplify turbulent flow, the complex turbulent eddy motion is replaced by a single eddy size, the most energetic eddy as introduced by Townsend. Introducing these highly energetic eddies implies that they will cause the highest elongational strain rates, either between an eddy pair, or during injection of fluid in the turbulent flow, and therefore be responsible for the onset of the polymer action (Durst 1982).

With the average elongational strain rate of the flow field between the eddies and the relaxation time of the polymer molecules, one can define a dimensionless characteristic number, the Deborah number, which represents the ratio of a characteristic time of flow and a characteristic time of the polymer molecule, and thus one can transfer considerations in porous media flow to the turbulent flow region.

The drag reduction is connected with a reduction in the burst frequency and structure, which is closely associated with the strong polymer-induced increase in elongational viscosity. Here, it must be taken into account that the elongation of the polymer

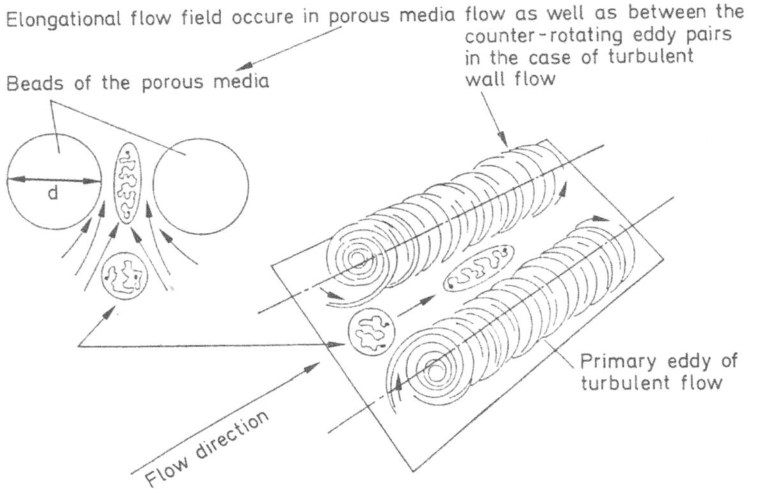

Fig. 6. Schematic representation of examples of the elongational flow between two eddies in turbulent flow in comparison with porous media flow (Durst 1982)

molecules does not take place during the burst directly from the equilibrium state. The polymer molecules are in fact already subjected to a shearfield in the proximity of the wall due to the shear flow already present before they experience the additional elongational flow (Bewersdorff 1986).

James (1987) and Wunderlich showed that prior shear had a considerable effect on the polymer solutions, causing significant extensional resistance at strain rates of the order of $10^2\ s^{-1}$, while rates one order higher were necessary to generate an extensional effect when there was no prior shear.

For several years, a controversy has existed as to whether the onset phenomenon could be defined by a length scale or a time scale (Lumley 1969, 1973). In turbulent flow near a wall it is possible to identify a smallest length and a least amount of time. Hence, the question arises whether onset occurs at some typical value of the ratio of length of time characterizing the polymer molecule. Such scales might be the root mean square radius of gyration of the undistorted molecule and the terminal relaxation time. The correlation between onset behavior and the Deborah number and the results of many experiments (e.g. Berman 1974; Gampert 1982) and theoretical considerations (Lumley) showed that the time scale hypothesis would appear to be relevant to the onset of drag reduction with polymers.

In contrast to this, Hlavacek took up once again the Virk linear measurement theory, whereby he posed the following considerations:

Virk and his colleagues at the time proceeded on the assumption that the eddies were larger than the macromolecules by a factor of 1.000. The size of the eddy diameter was at the time established as being of the order of $1 \cdot 10^{-2}$ mm. Hlavacek, proceeding from the study of eddy production, postulated a measurement of $1 \cdot 10^{-5}$ mm for the initial stages of the eddy. The inertia radii of high-molecular-weight polymers already lie in this region.

The proposed mechanisms of models to explain the drag reduction phenomenon are based on either a molecular approach or fluid dynamical continuum considerations, but these models are mainly empirical or semi-empirical in nature. Models constructed from the equations of motion (or energy) and from the constitutive equations of the dilute polymer solutions are generally not suitable for use in engineering applications due to the difficulty of placing numerical values on all the parameters. In the absence of a more generally accurate model, semi-empirical ones remain the most useful for applications.

5 Experiments, Methods, Apparatus

Using various test devices one can differentiate between two flow forms to detect the drag reduction phenomenon: the so-called "external flow" which includes flow over flat plates as well as around submerged bodies such as ship hulls of submarine models. The second form, named the "internal flow", includes flow situations in pipes, ducts, and pumps (Fig. 7).

One of the more important aspects of external flows is that the solid flow boundary is of finite extent (in the downstream direction) so that the flow must be treated as developing boundary layers in which time average steady-state conditions cannot be reached before the solid surface terminates. Another characteristic of such flows is

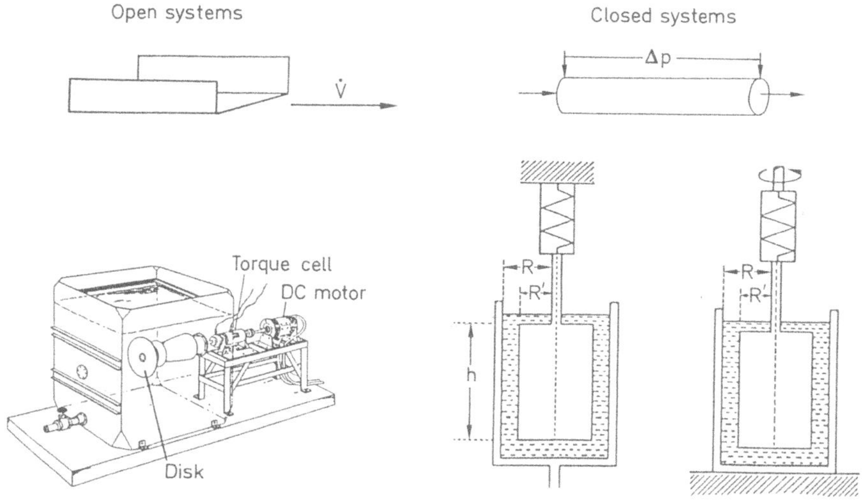

Fig. 7. Schematic representation of examples of test geometries used. Open systems: channel, rotating disc (Hoyt 1972, 1986). Closed systems: pipe flow, couette- or searle-systems (Kulicke 1986)

that they develop in a semi-infinite fluid body in which the growing boundary layer does not normally interact with those from neighboring surfaces. The external flow of polymer solutions on rotating disks and cylinders, and over flat plates offers an opportunity to examine hydrodynamic changes, this time in a boundary layer situation, produced by macromolecules. Thus a disk was rotated by an electric motor at such a speed that turbulent flow extended over the disk. Disk speed and torque were measured with various concentrations of polymer additives in a test tank. Either homogeneous polymer solution flows past the fixed plate or else water flows past the plate and a more concentrated polymer solution is injected into the developing boundary layer from a narrow slot in the plate close behind its front section. Another possibility is to tow the flat plate on a carriage through static water.

The torque of rotating cylinders is also reduced in polymer solutions under turbulent flow conditions. Analytical studies of the flow of polymer solutions over flat plates have all been based upon the similarity between flow in pipes and the flat plate flow.

In the case of "internal flows" extensive experimental data are available for turbulent pipe flow. The study of turbulent-friction coefficients in pipe flow has brought forth a number of effects displayed by flowing polymer solutions. Furthermore, many hydrodynamic investigations in pipe flow have been made to elucidate the flow behavior (laminar and turbulent) of Newtonian fluids. Thus, the pipe is one of the most investigated and traditional pieces of test apparatus and one can easily compare the flow behavior of Newtonian fluids and polymer solutions under constant boundary conditions.

The aim to build an optimal test device for our investigations has led to many helpful discussions with hydrodynamic scientists[1].

[1] The authors would like to express their appreciation to Prof. Durst, University of Erlangen, and Prof. Gampert and Dr. Wagner, University of Essen, for their helpful suggestions.

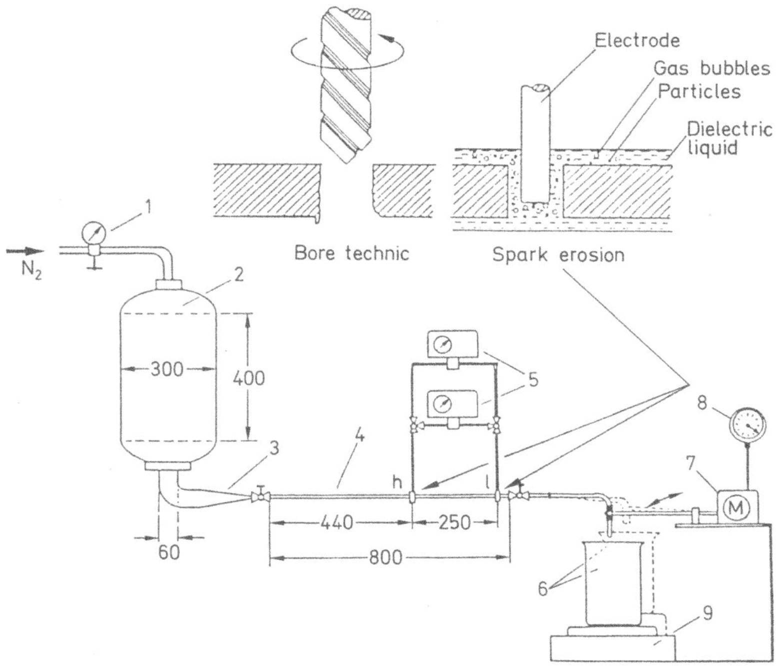

Fig. 8. Spark erosion technique and drag reduction equipment.
1 Precise manometer;
2 Fluid reservoir;
3 Cone;
4 Measuring length (250 mm);
5 Differential pressure transmitter;
6 Collecting reservoir;
7 Magnetic swivel arm;
8 Stop clock;
9 Balance

In order to examine the effectiveness in drag reduction of polymer solutions, it is necessary to employ a set of apparatus, which will help to test and determine the decrease in friction. Figure 8 illustrates the appropriate type of apparatus required, according to Eq. (6):

$$DR = 1 - \frac{f_a}{f_s} \tag{6}$$

and the equation (7):

$$f = \frac{2\,\Delta p \cdot d}{\varrho \cdot u_m^2 \cdot l} \tag{7}$$

The pressure drop along the pipe and the discharged volume must, in addition, be determined per unit time. The apparatus is designed in such a way that the mechanical strain on the polymer solutions is kept to a minimum. Consequently, pumps, sharp-edged quarter bends, and links have been left out. Apart from the mechanical degradation of the pump, obstacles to the flow, such as edges, sharp-edged links or ringblends and sudden deviations, for instance, as in the case for of magnetic valves, can also cause molecules to shatter. In order to keep these external influences to a minimum, the links and valves have to be built so as not to hinder the flow in any way. Moreover, ball valves have to be used, which do not produce a greater loss in pressure than with a flow-through pipe with the same diameter and length.

Instead of a pump, a gaseous medium, namely nitrogen from a gas cylinder was used as an accelerator, so that additional molecular strain in the form of shear stress, as is the case when using pumps, will not occur. With the aid of a high precision balance, it is possible to determine the volume by weighing the discharged volume. In order to avoid secondary flow influences, it is necessary to use quarter bends with a radius equal to or greater than $10 \times d$ (diameter). The strain on the molecules is kept to a minimum by using a connecting quarter bend on the storage tank, which has a relatively large internal diameter. Since the measurements themselves only last a few seconds, the possibility of error during the outflow of the test fluid, in the form of lowering the liquid level in the storage tank, is so minimal that it is negligible. The diameter of the tank should be relative to the height of the tank. There should be at least 100 ml of fluid flowing through the equipment for each point of measurement, so that errors can be avoided. The nitrogen control valves are intended to keep the pressure above the solution in the storage tank and during the measurement at a constant level.

In order to obtain a higher level of control, two valves, one for less accurate and one for more precise regulation, were installed. One must ensure that in the test pipe itself, the measuring holes are equal or smaller than 1/10 of the internal diameter of the pipe. To avoid the formation of eddies or other influences on the flow around the measuring holes, which can lead to false results, the holes must not have any rough edges or roundings. In order to fulfill all these conditions, the measuring holes are produced with the aid of the spark erosion technique[1].

This technique has some advantages over the drilling method: the electrical sparks produce craters or little cavities, whereby mechanical processing leaves behind grinding-, rotator-, or milling-groovers. The crater dimensions depend on amperage, voltage, capacity, and the ratio of charging and discharging times. A quick spark sequence and a small working speed provide only a little roughness (0.001–0.005 mm depending on the material). Another advantage would be to produce a hole with an angle of 90° without sharp or rounded edges.

The opening of the test pipe has been designed and tested to ensure fully developed turbulent flow in the test section. Hence, the inlet length of the pipe to the first pressure gap amounts to $220 \times d$. The device has been carfully tested and calibrated with water, which complies with the Newtonian theory of fluids for smooth pipes. Measurements are taken after stabilization occurs and then a magnetic device in the form of a swinging arm turns the outlet tube to the collecting vessel and at this moment a stop-watch is

[1] At this point we want to thank Dr. Drews from the PTB, Braunschweig for his generous support in applying the spark erosion technique.

started, and stopped when the measurement finishes. So one can estimate the flow rate by measuring the weight of the liquid and the stopped time. All parts of the test device which are in contact with the test liquid are made of stainless steel to avoid corrosion effects, especially the formation of rust particles which can interfere considerably with the measurements.

Among the great number of publications, there are only a few reports in which the influence of different capillary or pipe material is investigated. Tests have been made mainly with stainless steel (Hershey) or glass (Dschagarowa 1980), bronce (Toms, Nagarajan), teflon (Peyser), brass (Gold), and copper (Dschagarowa 1980). The influence of the type of capillary material on the drag reduction of toluene solutions of polyisobutylene (PIB) with different molecular weight was investigated in the work of Dschagarowa (1978). Capillaries of steel, copper, and glass with diameters of

Table 2. Flow improving additives used to date

Solid medium	Fibrous medium
Fine-grained sands	Wood
Small glass beads	Nylon
Flax grains	Peat
Coal dust	Rayon
Lubricating grease, corundum	Asbestos, glass
Micelle-forming system	
Tri-*n*-butyl-tin(IV) fluoride, cetyltrimethylammoniumbromide, cationic, anionic, and non-ionic surfactants	
Sythetic, organic polymers	
Polymethylmethacrylate (PMMA)	Polyethyleneoxide (PEO)
Polyisodecylmethacrylate	Polystyrene
Polyacrylic acid (PAA)	Polystyrenesulfonate
Polyacrylamide (PAAm)	Polyethylenimine (PEI)
Hydrolyzed polyacrylamide	Polyvinylalcohol (PVA)
Glyoxylyzed polyacrylamide	Polyvinylpyrrolidone (PVP)
Polyisobutylene (PIB)	Poly-*cis*-isoprene (PCIP)
Synthetic, inorganic polymers	
Polydimethylsiloxane (PDMS)	Polyphosphate
Biological additives	
Guar gum (GG)	Guarantriacetat (GTA)
Carboxymethylcellulose (CMC)	Hydroxyethylcellulose (HEC)
Xanthan	Fish mucus
Chlorella stigmataphara	*Porphyridium aerugineum*
Porphyridium cruentum	*Chaetoceros didynuis*
Protocentrum micans	*Arthrobacter viscosus*
Chaetoceros affinis	*Exuviella casszbuca*
Anabaena flos-aquae	*Chlamydenas peterfii*
Pseudomonas and *Neisseria* species	

2.00, 2.42, and 3.00 mm, respectively, were used. It was found that drag reduction is not affected substantially by the type of capillary material, although the friction coefficients of the solvent, calculated from the experimental data are different according to the sequence: steel > copper > glass. The friction coefficients of the investigated solutions are arranged in the same order.

6 Flow Improvers in Use

A number of different additives have shown themselves to be successful drag reducers.

Experiments not only in organic but also in aqueous dispersions were carried out in order to elucidate the drag reduction phenomenon. It is possible to divide the flow improvers used into three main groups:

1) Solid particles (sands, fibers);
2) colloidal systems (surfactants, soaps);
3) polymers (synthetic polymers and biopolymers).

6.1 Solid Particles

The "solid particles"-group consists of suspensions of different types of sands, rubber granulates, as well as natural and synthetic fibers (e.g., see Reddy 1986). Turbulent drag reduction studies were carried out in the same way on both glass and sand in poly(ethyleneoxide) solutions (Ishibashi) and on asbestos fiber mixtures with biopolymers (e.g., guar gum, xanthan, graft copolymers) (Malhotra; Reddy 1985). Research into the drag reduction effect was initiated at the beginning of this century when, by coincidence, it was discovered that the friction factor during the hydraulic transportation of these additives was reduced. Hence, it was possible to ascertain that cloudy fluids flow faster than clear ones. It was, however, not until the 1930's and 1940's that intensive research into this phenomenon was started. It was found that as far as fibers were concerned, those with a large length-diameter ratio were especially effective. Although solid particles have not been studied as thoroughly as polymers and surfactants, there are nevertheless two important factors when they are put to industrial use: it is easy to add solid particles to the flow medium and then withdraw them again later. They are also not prone to mechanical degradation. However, the degree of effectiveness achieved, in comparison to polymers and surfactants, is not so great.

6.2 Micellar Systems

Micellar systems (i.e.,. Shenoy 1984; Ohlendorf; Brunn) as well as other colloidal systems (polyphosphates (Hunston), tri-*n*-butyl-tin-fluoride, e.g. Dunn; Evans) come under the heading "surfactants". It is necessary to differentiate soaps into anionic, cationic, and non-ionic types. Among the anionic types one can find, for instance, alkali metals and ammonium salts consisting of various fatty acids, which were studied in aqueous and non-aqueous systems. They are able to form micelles which even after being destroyed are still capable of regenerating themselves with diminishing stress. The soaps indicate a high degree of long-term stability. Anionic agents are not expensive, but since they have to be applied in an aqueous system, their usage is limited. When calcium and other cations are present, precipitation will occur. An interesting aspect of anionic soaps in aqueous and organic solvents is that they demonstrate contrasting behavior at different temperatures. In non-aqueous solvents they were particularly effective at lower temperatures.

In cationic soaps it becomes evident that cetyltrimethyl-ammoniumbromide (CTAB) in aqueous solutions with naphthaline derivates produces viscoelastic gels. This CTAB-naphthol mixture behaves like a concentrated drag-reducing fluid and therefore does not demonstrate transitional behavior when changing from the laminar to turbulent stage. If sand-water suspensions are added to this mixture, it is also possible to reduce the friction factor. Small-angle-neutron-scattering (SANS) measurements, which were done with a dilute solution of *n*-tetradecyltrimethylammonium bromide (TTAB) with sodium-salicylate in pipe and channel flow show that the micelles exist above the critical wall shear stress (Bewersdorf, Frings 1986). Cationic surfactants display an important advantage over anionic surfactants in that they do not precipitate in the presence of calcium ions. On the other hand, they are more expensive, and in an aqueous state, there is evidence of chemical degradation after several days. Although mechanically stable, they are thermally instable thus limiting their range of application.

Studies on non-ionic surfactants as effective drag-reducing additives have been submitted by Zakin (1972). He studied various effects on three non-ionic surfactants formed from straight-chain alcohols and ethyleneoxide. These surfactants have an upper and a lower temperature limit for solubility in water and prove effective drag reducers near their upper critical solubility temperature or clouding point. The clouding point is the point at which a solution of a non-ionic agent in water becomes turbid as the temperature is raised.

The non-ionic surfactants have an advantage over all the drag-reducing additives studied so far. They are both mechanically and chemically stable, do not precipitate in the presence of calcium, and hence can be used in all impure waters, sea water, brackish water, or concentrated brine solutions. They have an excellent potential for drag reduction at high temperatures.

6.3 Polymers

In the third group of flow improvers one finds the synthetic and biopolymers. In comparison to the other drag reducers, they have proven to be the most effective, and among the polymers themselves the synthetic ones are more effective than the biopolymers (Fig. 9).

It must, however, be borne in mind that a comparison such as that in Fig. 9 is incomplete, because nothing is known about the molecular parameters. As already stated in the summary, it is the aim of the present investigations to research the comparative influence of molecular parameters of polymers with respect to their effectiveness as drag reduction factors. The aforementioned molecular parameters include: molecular

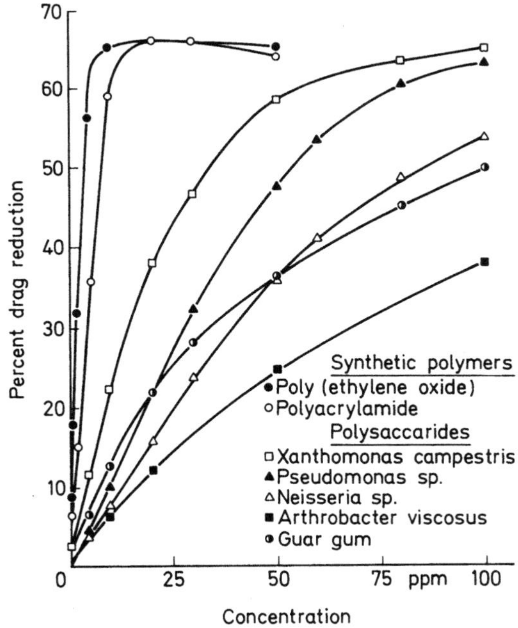

Fig. 9. Friction reduction effectiveness of synthetic polymers and polysaccharides during one pass through a turbulent flow rheometer (data taken from Kenis)

weight, molecular weight distribution, chemical nature of the polymer, structural parameters, solvent quality, etc.

Within this group, the linear polymers have been most intensively studied and researched. The main area of interest lies in water-soluble polymers. However, there have also been many studies into organic solutions, although there exist problems with odor and toxicity in the case of analytical studies. Among the water-soluble additives, tests were made especially on polyethyleneoxide (PEO), polyacrylamide and the coacrylates (PAAm, PAAm/AAcNa), polyacrylic acid (PAA), guar gum (GG), carboxymethylcellulose (CMC) and sodium salts, as well as hydroxyethylcellulose (HEC). From these tests, PEO proved to be the most effective flow improver, followed closely by PAAm, which is somewhat more stable than PEO in turbulent flow.

In organic solvents polymethylmethacrylate (PMMA), polystyrene (PS), polyisobutylene (PIB), and also polyethyleneoxide (PEO) were mainly studied. Almost all of the polymers used in the experiments are industrial products whereby most of the results available deal with PEO and PAAm. Apart from single polymers, polymer mixtures have also been tested (Dschagarowa 1978), for instance, PIB and PS in toluene as well as polyethyleneimine (PEI) with PEO, PAAm, and PAA. Experiments have also been carried out using interpolymer complexes formed by one polymer with anionic and one polymer with cationic groups (Kowalik), polymers and dyes (Stenberg), and polysaccharides-based (Guar Gum, Xanthangum) graft copolymers (Deshmukh 1985, 1986, 1987; Rochefort). The great advantage of polymers is their effectiveness at very low concentrations; however, they are prone to mechanical and thermal degradation.

6.3.1 Problems in the Characterization of Polymers — Molecular Weight, Molecular Weight Distribution, Maximum Behavior, Critical Concentrations

Many experiments have shown that polymer additives are the most effective drag-reducing agents of flow improvers, and great efforts have been made in the hydrodynamic region to elucidate the drag reduction phenomenon. But investigations into the polymers themselves are relatively rare, especially in the case of highly diluted solutions which are employed in drag reduction experiments. In nearly all investigations, commercial polymer products were used. In such cases a precise characterization, in terms of molecular weight and molecular weight distribution, was not performed and therefore a comparison of experimental data is not possible. Our intention is to identify the molecular parameters, which are of particular relevance for the qualitative understanding of the flow phenomena caused by water-soluble polymer additives. The parameters can be classified with respect to *primary*, *secondary*, *tertiary*, and *quarternary* structure. The primary molecular structure parameters comprise the molecular weight in general ($M = 10^5$–10^8 g/mol) and the chemical nature of the polymer (e.g., ionic or non-ionic agents). The secondary molecular structural parameters are the molecular weight distribution, branching and copolymer composition. In the case of the tertiary molecular structure, parameters like molecular shape, radius of gyration, critical concentration, and the second virial coefficient exist. The quarternary structure parameters comprise dynamic parameters in general, interactions (segment mobility), entanglements and energetic interactions (association, aggregation, and hydrogen bonds). A reason for the problem encountered in investigating highly diluted water-soluble polymer solutions can be seen in the fact that the purification of aqueous solutions is far more difficult than in the case of organic solvents. This is due to the polar character of water, which attracts dust, fibers, and other particles more readily. In such cases the determination of the microstructures, using light-scattering parameters is hindered and very often impossible, thus, one has to improve the purification procedure.

The determination of the absolute average molecular weight, which refers to the *primary structure* parameters, should be discussed first, because it is mainly responsible for the viscoelastic flow phenomena observed with polymer solutions in shear and elongational flow. The absolute average molecular weight may be determined, either by light scattering (M_w) or by osmotic pressure measurements (M_n). It is a fact that with increasing molecular weight, polymer additives show pronounced flow phenomena, especially in the cases of drag reduction and porous media flow. Ultra-high molecular weights can only be measured by using the light-scattering technique. Osmotic pressure measurements are restricted to molecular weights up to $1 \cdot 10^6$ (g/mol). As mentioned above, the contamination of polymer solutions by dust and fibers is a major problem in light scattering. Therefore, one has to purify the solutions very carefully, which is often not easy, especially when commercial polymer samples are used. An easily manageable purification arrangement is shown in Fig. 10a. The solution in the store tank is filtered through a membrane which is placed in a stainless-steel syringe holder into the dust-free scattering cell (Kulicke 1980).

Using this arrangement, absolute clean scattering cells are not required because extraneous particles are removed by continued back pumping.

When the scattering cell has been purified with accuracy, the arrangement can be simplified by directly connecting a gas-tight syringe with a filtration holder to the scattering cell (Fig. 10).

From low-angle-laser-light-scattering measurements, one can take the absolute average molecular weight and the second virial coefficient A_2, which is an indication

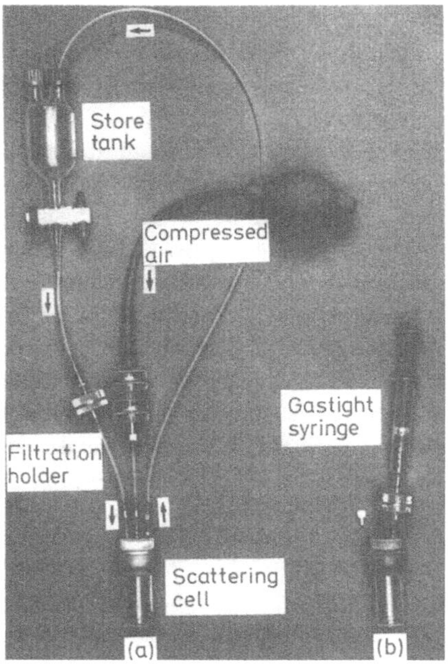

Fig. 10. Purification of polymer solutions.
a) Filtration equipment;
b) Gastight syringe

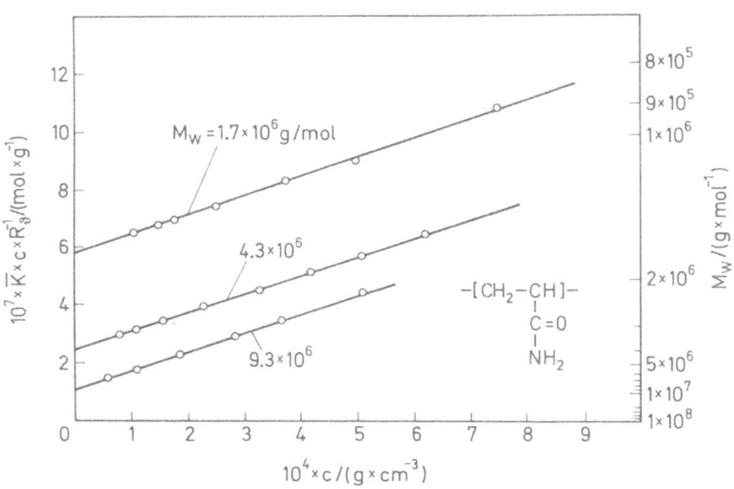

Fig. 11. Low-angle-laser-light-scattering results of polyacrylamides in 0.1 M Na_2SO_4-solution at 25 °C. Note at $K_\theta \cdot c/R_\theta = 0$ the molecular weight should be $M_w = \infty$ definition

of polymer-solvent interactions. Theoretically, it is possible to measure molecular weights up to infinity with the use of light scattering. The point of extrapolation of the straight line of the experimental data with the ordinate at $c = 0$ gives $1/M_w$. At $K_\theta \cdot c/R_\theta = 0$, the value of M_w goes to infinity. The corresponding molecular values are plotted to the right of this figure. Therefore, above a molecular weight of $5 \cdot 10^6$ g/mol, the possibility of error becomes very significant. The reason is that the molecular weight M_w increases hyperbolically as $K_\theta \cdot c/R_\theta$ decreases, as can be seen by the right ordinate of Fig. 11.

Therefore, the determination of ultra-high molecular weight polymers is, in general, not easy to perform (purification problem), and in addition the determination of increasing molecular weight becomes increasingly limited.

Other absolute methods for detecting high molecular weights lead to similar problems. Therefore, the normal method for the determination of these high molecular weights is by extrapolation of the Mark-Houwink relationship $[\eta] = K \cdot M^a$, where K and a depend on the selected polymer-solvent-temperature system. This Mark-Houwink relationship (found in polymer handbooks, e.g., Brandrup, Immergut) is limited in most cases to the determination of absolute molecular weights lower than $2 \cdot 10^6$ g/mol. The advantage of this method is that the reduced viscosity can be determined quickly, which is, moreover, much cheaper and less time-consuming than absolute determination methods such as light scattering measurements. The value of η_{sp}/c at $c = 0$ gives the intrinsic viscosity $[\eta]$, from which the molecular weight can

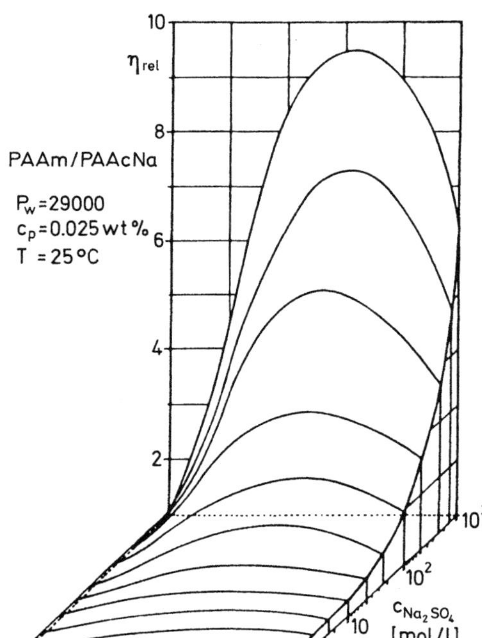

Fig. 12. The relative viscosity of polyacrylamide-co-acrylates versus different copolymer compositions depending on the presence of the low-molecular-weight electrolyte sodium sulfate. All Samples have the same degree and distribution of polymerization

be easily calculated. An increase in the intrinsic viscosity reflects a higher molecular weight of the polymer sample. The higher-molecular-weight samples form a slight curve. This is due to intermolecular interaction between polymer molecules. The determination of the intrinsic viscosity should be performed in the dilute state of the solution. However, in the dilute concentration range, the increase in the polymer solution viscosity compared to the solvent is too small and therefore most of the points of measurement are performed using a moderately concentrated solution. This is one of the disadvantages of using relative determination methods, especially with increasing molecular weights (Kulicke 1986).

In the case of polyelectrolytes, another parameter has to be considered: the degree of dissociation. Molecular weight determination in pure water can not be performed. The reason for this is that the viscosity is also very high in the dilute state and therefore the purification procedure (filtration, centrifugation) is difficult to perform. In addition, the difference between the light scattering intensity in the solvent and in the polymer solution is very small at this low concentration, which makes such measurements impossible. Therefore, all measurements have to be performed in a low-molecular-weight electrolyte. In Fig. 12 one can see that by increasing the content of charged side groups, the viscosity at a given salt concentration increases, but it is possible to continuously suppress the Coulomb repulsion with increasing salt concentration.

Therefore, when dealing with polyelectrolytes, countless Mark-Houwink relationships must be established. This shows the increasing problems of simple molecular weight determinations on charged systems (for more details, see Kulicke. Hörl (1985)). Thus, the molecular weight determination of polymer molecules or additives is an important fact in the drag reduction area. Molecules or particles which are effective have molecular weights above 10^5 g/mol (the polymer backbone chains should be linear, flexible, and unbranched).

The influence of the molecular weight distribution, which refers to the *secondary structure parameters*, on the behavior of solutions as well as the method of determination will now be discussed. By preparing two polyethyleneoxide samples with different molecular weight distributions (M_w/M_n = 1.6 and 4.1) but the same average molecular weight, it was found that the zero-shear viscosity differs by a factor of ten. This also shows clearly that every Mark-Houwink relationship depends on the quality of the homologous series of polymer samples, where every sample has exactly the same molecular weight distribution. This is obviously impossible, because the preparation of different molecular weights involves at least a small change in the molecular weight distribution. Therefore, in the last ten years increasing interest has been focused on the determination of molecular weight distributions. The normal procedure in the determination of the mass of particles involves the use of the refractive index, whereas the separation of the polydisperse sample is caused by using the size-exclusion-chromatographic method. To calculate the molecular weight distribution, one has to know the calibration curve for the polymer-solvent system. This is in fact an indirect method and, in addition, as in the case of aqueous polymer solutions, some peculiarities may be observed. The molecular weight distribution, including the calculation of M_w and M_n for an aqueous polymer solution system, was recently determined by simultaneously and continuously monitoring the effluent from a size-exclusion-chromatographic column (SEC) filled with porous glass beads with a low-angle-*laser*-light-scattering (LALLS) photometer; see Fig. 13 (Kulicke, Böse 1984).

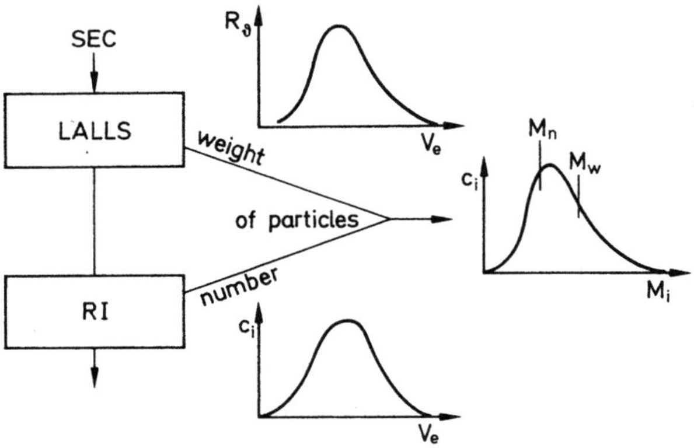

Fig. 13. Schematic representation of the absolute determination of the molecular weight distribution of aqueous polymer solutions, using a combined size-exclusion-chromatographic and low-angle-light-scattering apparatus

The recorder traces of the scattering intensity and the refractive index difference have been digitalized in about 400 steps corresponding to 200 μl fractions. Each fraction has been assumed to be monodisperse. This method gives absolute data rather than relative data obtained from all other calibration methods. The investigations clearly show a flow-rate-dependent degradation of samples with a relatively low molar mass ($M_w = 5 \cdot 10^5$ g/mol, $M_w/M_n = 2.5$) at typical flow rates Q = 1 ml/min. Consequently, such a determination of the absolute molecular weight distribution is not an easy task. The characteristic ratio M_w/M_n has often been used to indicate the width of the molecular weight distribution, which is, of course, the initial stage of approximation and can also only be used up to an average molecular weight of $1 \cdot 10^6$ g/mol. On the other hand, the high-molecular-weight tail of the distribution function subsequently limits this method. To overcome these shortcomings, one has to improve this method at least for higher-molecular-weight samples by changing matrix, volume speed, thermodynamic quality of the solvent, etc.

The drag reduction effect is very sensitive to molecular weight, especially to the chain length. Hence, the higher the molecular weight, the better the friction reduction and together with the polymer concentration it can be lowered. Wagner has shown in his thesis that the high-molecular-weight part of the molecular weight distribution defines the drag reduction and the onset behavior of polymers.

In general, good polymer drag reducers should have a linear and flexible backbone chain without branching. To ensure that we get such polymers, we have prepared polyacrylamide in our laboratory. Polymerization with H_2O_2 has the obvious advantage of producing linear, completely soluble polymers if the conversion is kept below 20%. On the other hand, no residual initiator or ions can be present in the polymer (Kulicke 1982). In solution the polymer molecules are usually randomly coiled so that they form tangled and beanlike coils filled with solvent. Good solvents interact with the macromolecules causing an expansion of the coil. Another constellation is possible in the case of polyelectrolytes. In solution they show an expanded, more rodlike shape which is caused by the repulsion of the similar charged ionic side groups. This extension is very sensitive to the pH of the solution. Thus, it is possible to change the coil conformation by introducing ionic side groups into the backbone chain. This modification in the polymer shape can be measured with a wide-angle-light-scattering photometer by determinating the radius of gyration.

Therefore, we take our own preparations of polyacrylamides, produced in free radical polymerization in aqueous solutions with the molecular weight distribution of $M_w/M_n = 2.5$, and with the aid of alkaline hydrolysis we can introduce acrylate side groups to various contents depending on the reaction time. This saponification with aqueous sodium hydroxide at room temperature leads to the sodium salts (polyacrylamide-co-acrylates) which have the same degree of polymerization and the same molecular weight distribution as basic polyacrylamide. This is a so-called "polymer analogous reaction". In order to attain an hydrolysis level below 30 mol-% acrylamide (AAm) (this value is the limit under alkaline conditions), one can take the highest saponified product from the alkaline saponification and treat the sample under acidic conditions providing higher saponification degrees. In this way we get homologous series of polymer samples which are comparable and it is possible to examine the influence of the coil volume when chain length and molecular weight distribution are constant; see Table 3, 4, and 5.

Table 3. Molecular data of the non-ionic poly(acrylamides) used

Sample	M_w (LS) (g/mol)	M_w (LALLS) (g/mol)		$[\eta]$ (cm^3/g)
PAAm 37/05	$13.7 \cdot 10^6$	$10.1 \cdot 10^6$		1577
PAAm 40/1	$8.2 \cdot 10^6$	$8.2 \cdot 10^6$		1405
PAAm M 3*	$3.9 \cdot 10^6$	—		797
PAAm B 1*	$1.1 \cdot 10^6$	—		329
	M_η (g/mol)	d_{equ} (nm)	$\langle R_G^2\rangle^{1/2}$ (nm)	A_2 ($bar \cdot cm^6 \cdot g^{-2}$)
PAAm 37/05	$10.3 \cdot 10^6$	274	314	3.0
PAAm 40/1	$8.8 \cdot 10^6$	250	230	6.55
PAAm M 3*	—	158	173	7.31
PAAm B 1*	—	77	86	7.52

Table 4. Molecular data of the poly(acrylamide-co acrylate) used

Sample	M_w (calculated) (g/mol)	$[\eta]$ (cm^3/g)	d_{equ} (nm)
COP 40/1/1	$8.5 \cdot 10^6$	2097	283
COP 40/1/4	$9.0 \cdot 10^6$	2952	322
COP 40/1/6	$9.9 \cdot 10^6$	4505	383
COP 37/05/1	$10.4 \cdot 10^6$	1782	286
COP 37/05/4	$11.0 \cdot 10^6$	3319	358
COP 37/05/6	$12.2 \cdot 10^6$	4474	410

Table 5. Determination of the copolymer-composition of the poly(acrylamide-co acrylate) used

Sample	Part of acrylate (mol %)			Residual water content (weight %)
	^{13}C-NMR	IR	EA	
PAAm 40/1	—	—	—	4.6
COP 40/1/1	13	14	13	2.6
COP 40/1/4	28	34	30	5.2
COP 40/1/6	61	58	64	7.1
PAAm 37/05	—	—	—	6.1
COP 37/05/1	—	10	10	2.9
COP 37/05/4	—	26	28	3.9
COP 37/05/6	—	60	64	8.6

Ting (1972) reported that the degree of drag reduction depends on the content of ionic groups in a series of hydrolyzed PAAm samples exhibiting "maximum behavior". He noticed that the reduced viscosity increased sharply with increasing hydrolysis up to 40% AAm and then decreased, in comparison to the pure polyacrylic acid. Similar investigations with polyacrylic acid at different neutralization states have been made by Parker and Banijamali who obtained comparable results.

Klein performed a more detailed investigation on the synthesis and molecular structure of a homologous series of PAAm/AAcNa with an AAm content between 33 mol-% and 100 mol-%, which had the same level of polymerization and the same chain length distribution. He found that the intrinsic viscosity as well as the exponent of the Mark-Houwink equation rises with increasing AAcNa content, even when the measurements were performed in salt solutions. These investigations are restricted to the alkaline hydrolyzed polymer samples up to an AAm content of 33 mol-%. The findings of Kulicke, Hörl confirm the predicted results that the reduced viscosity of coacrylates with an AAm content below 30 mol-% decreased. Similar to the viscosity, the radii of gyration of the polymer samples show the same maximum behavior. In addition, this so-called "maximum behavior" is not observed in physical mixtures of PAAm and PAAcNa. Measurements of viscosity and light scattering have proven that maximum coil dimensions of PAAm/AAcNa in salt solutions are achieved at about 67 mol-% AAcNa. The maximum behavior was researched by measuring the radius of gyration, second virial coefficient, molecular weight, preferential solvation, absorption band, and viscosity of the copolymer series (Kulicke, Hörl 1985). Thus, a decrease in the possible ways of arranging AAcNa-AAm-AAcNa-units (via H-bonds) will lead to extended coil dimensions and therefore influence the viscosity level. In addition, it can be concluded from the results that the maximum behavior is real and not influenced by the different distributions of the samples.

All factors related to the arrangement of the polymer chain in space are classified as *tertiary structure*. Parameters measurable directly (the radius of gyration R_G, the end-to end distance h, the hydrodynamical radius R_H, and the asymmetry in light scattering intensity) or indirectly (interaction parameters, the second virial coefficient A_2) are related to the dimensions, such as size and shape of the polymer chain in a specific solvent under given conditions of temperature and pressure. For the exact determination of the coil size of macromolecules, it is necessary to ensure that measure-

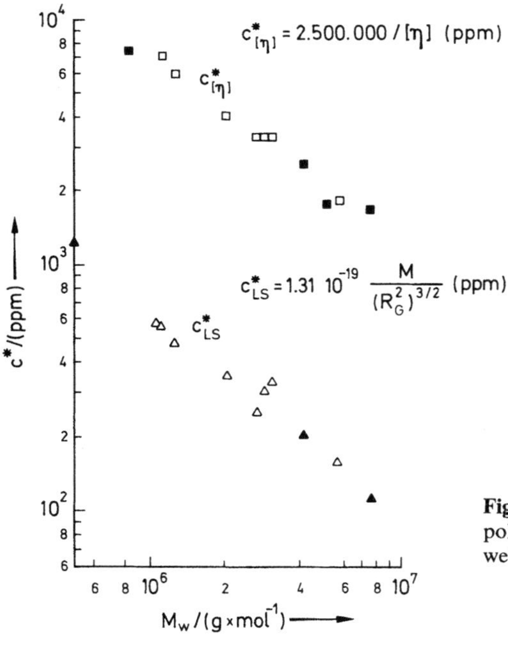

Fig. 14. Critical concentrations c* of non-ionic polyacrylamide in water at 25 °C versus molecular weight $M_w/M_n = 2.5$

ments are carried out in dilute solutions, because in moderately concentrated solutions intermolecular interactions will lead to errors in the results. In general, the range of dilute particle solutions and concentrated solutions may be definitely separated by a critical concentration c*. This can easily be determined by using viscometric measurements which yield a relative value, or by performing, for example, light scattering measurements, which yield absolute values of the critical concentration. However, the viscometric value of the critical concentration (cf. Sect. 6.3.3.3) is not equivalent to the absolute one of light scattering measurements because only absolute methods like light scattering give correct values of c*. Figure 14 demonstrates the results of such investigations. One can see that both methods lead to a linear dependence between c* and the molecular weight M_w but differ by a factor of ten.

Similar results were not only found for polyacrylamide in water but also for polystyrene in toluene and polystyrene in *cis*-decalin, which show that this was not an exception. In both cases, the geometric shape of the polymer coils are assumed to be spherical, but in accordance with the findings of Kuhn and others we know that the most probable form can be best represented as a bean-like (irregularly ellipsoidal) structure. A polymer chain in dilute solution can be seen as a coil which changes its form continuously with time under the influence of statistical thermal movements and the bean-like geometry includes the largest number of possible formations (random flight statistic). An important fact is the thermodynamic quality of the solvent, because polymeric flow improvers should be easily soluble and good solvents cause an expansion of the polymer coil and thus the flexibility of the polymer chain is increased. In many investigations it was found that degradation of polymer molecules occurs earlier in bad solvents than in good ones, because the coil becomes more compact and rigid, and therefore the macromolecules are more sensitive to degradation (Brostow 1983, 1984).

From all these results, we came to the conclusion that the size of polymer molecules in solution is of key importance in increasing the effectiveness of technical applications. In a comprehensive study it has been shown that the coil dimensions can be influenced by ionic (anionic or cationic) side groups which will lead to an increase in the coulomb repulsion. Other methods can be used by changing the solvation or by introducing steric side groups. The last example has a very strong influence on the chain flexibility which can be measured in different ways (Kulicke 1982). However, these examples should only show that, besides the molecular weight and distribution, etc., the coil dimensions can be influenced by changing the chemical structure or, especially in the case of charged copolymers, by different salt concentrations in the solvent. This is comparable with investigations on polyacrylic acid in different neutralization states.

6.3.2 Polymer Handling and Aging

The polymer molecules depend very often upon the preparation conditions (polymerization and storing methods) and their handling. In moderately concentrated solutions, energetic interactions between the polymer molecules and also formation of aggregates or associations which are related to the quaternary structure parameters can occur. It has been frequently observed that aqueous PAAm solutions show a time-dependent viscosity, which has been investigated by several scientists but interpreted in different ways. The effect expresses itself in a long-term decrease of solution and intrinsic viscosity over the course of several weeks, even when the solutions are kept at room temperature, in the dark, and without shaking (mechanical influence). This unusual viscosity-time dependence is called solution instability and detailed investigations thereof have been carried out in our laboratory (Kulicke 1982, 1986). The following results were obtained:

The viscosity decrease did not only occur in technical products, but also in high-purified PAAm if the molecular weight exceeds $1.5 \cdot 10^6$ g/mol.

Light scattering measurements yield unchanged molecular weights; that means, no chain degradation occurs. The hydrodynamic volume becomes smaller with time

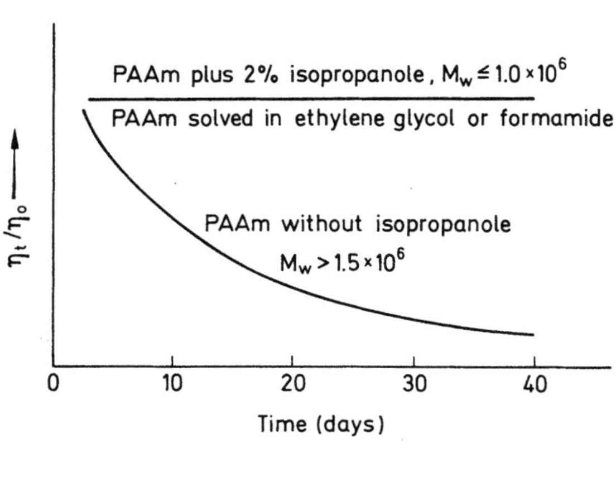

Fig. 15. Schematic representation of the time dependent solution viscosity of PAAm.
η_0 = viscosity at 1st day;
η_t = viscosity at tth day

as confirmed by SEC measurements and flocculation tests. The possible existence of hydrogen bonds in aqueous PAAm solutions, as an influence on solution instability, was supported by the addition of structure-breaking or -forming additives, and by infrared spectroscopy investigations on deuterated and undeuterated PAAm. The standardization of several pH values from 1 to 12 by means of buffer substances as well as the addition of hydroquinone, *p*-benzo-quinone, and Na_2SO_3 did not result in stable solutions. No viscosity loss could be observed in formamide and ethylene glycol, at low temperature (239 K) in water, or by adding 2 vol-% 2-propanol to aqueous solutions; see Fig. 15.

In the case of aqueous polyvinylalcohol solutions, a viscosity increase with storage time was observed (Fig. 16).

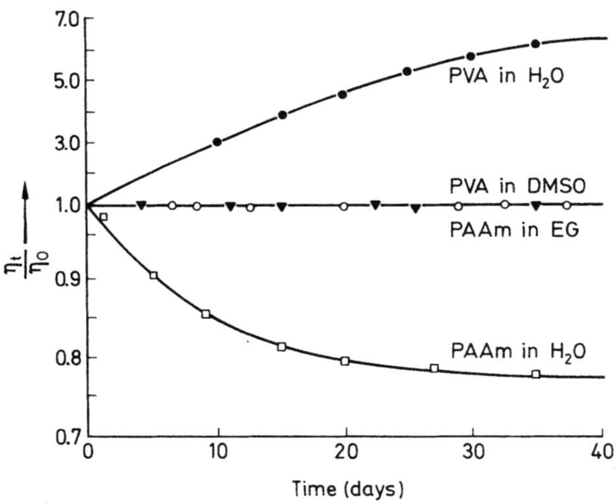

Fig. 16. The solution viscosity of polyvinylalcohol and polyacrylamide in different solvents as a function of storage time

Using dynamic light-scattering techniques, Lechner was able to prove that this effect is due to the tendency of PVA to agglomerate; for more details see Kulicke, Böse, Bouldin. On the other hand, a decrease in viscosity has been obtained with no detectable change in the average molecular weight.

The solution viscosity of aqueous polyacrylamide-coacrylates decreases drastically within several weeks (Fig. 17). The behavior is similar to that of PAAm.

When storing the solid samples under the possible influence of moisture, the intrinsic viscosity, even after intensive drying, was significantly smaller than directly after polymerization.

From these experimental results it was assumed that conformational transitions within single polymer chains — controlled via intramolecular hydrogen bonds — are causing the viscosity decrease. Chain scission could be excluded.

To explain the mechanism of the change in conformation, a model was suggested which is shown in Fig. 18.

Mainly extended segments with hydrogen bonds of type x and/or helical structures of type y exist in the beginning. The hydrogen bonds of type x are most effective if

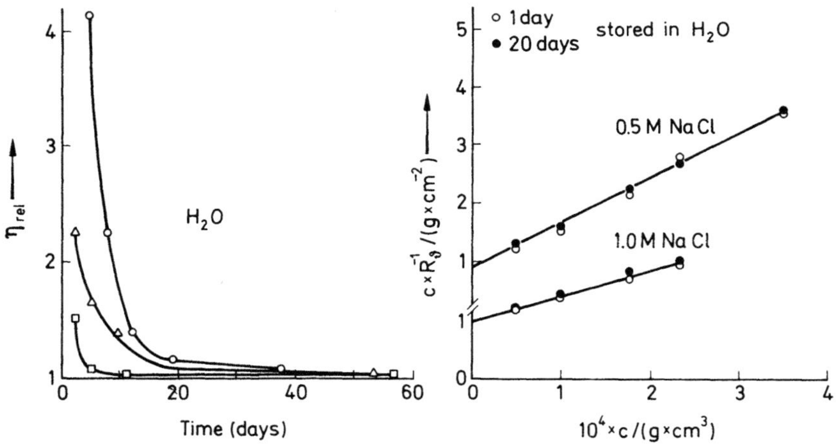

Fig. 17. Relative viscosity of aqueous PAAm/AAcNa-solutions (○: 58.5 mol-% AAm, △: 82 mol-% AAm, □: 92 mol-% AAm) dependent on storage time. c = $3 \cdot 10^{-6}$ g/cm^3 and light scattering results from measurements on the 1st and 20th day in two different salt concentrations (λ_0 = 633 nm, angle 6–7%, copolymer with 56 mol-% AAm)

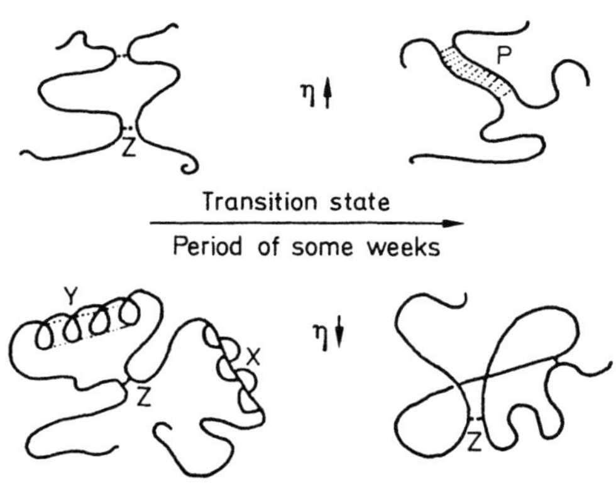

Fig. 18. The possible existence of hydrogen bonds in aqueous PAAm solution, as an influence on solution instability, was supported by the addition of structure-breading or -forming additives and by IR investigations on deuterated and undeuterated PAAm

the angle is about 180°, which can be demonstrated by means of molecular models. This demand is well realized for a 1–3 and 1–4 connection of side groups by hydrogen bridges. One may argue that water molecules can attack the hydrogen bonds of type x and y and that these oriented structures can be destroyed by cleavage. The oriented segments are mainly destroyed and a more flexible coil structure having a smaller hydrodynamic volume is formed.

Gampert also noticed a decrease in viscosity with time in aqueous PAAm solutions, but this viscosity loss does not influence the drag reduction effectiveness in his measurements. In contrast, Chang's studies showed that the rheological properties are time dependent and their drag reduction is reduced with aging. Hence, the drag reduction effect is clearly related to the chain length, and we can conclude that no chain scission has appeared (Fig. 19).

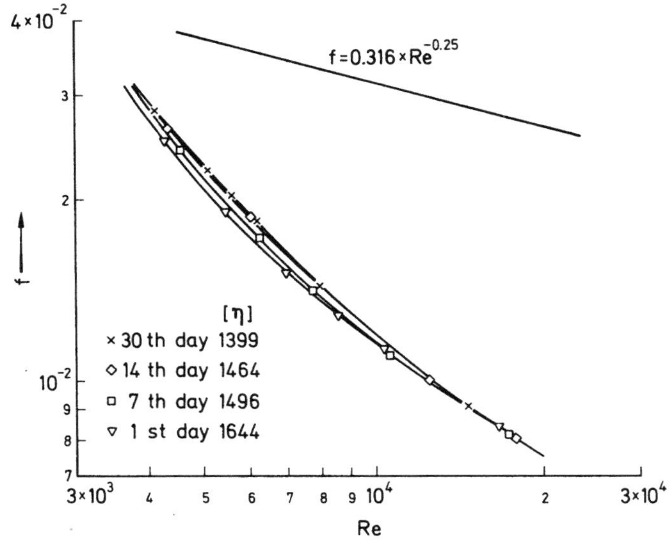

Fig. 19. Comparison of the intrinsic viscosity and the drag reduction material function (friction factor versus Reynolds number) dependent on storage time, $M_w = 8.1 \cdot 10^6$ g/mol, c = 50 ppm in distilled water, d = 6.1 mm (Wagner)

6.3.3 Drag Reduction Results in Pipe Flow

Extensive studies have been carried out using commercially available polymers, which in most cases involve the problem of the unknown molecular parameters. It is of some interest, however, to compare the various trade products (see, for example, LECA).

The aim of this contribution is to investigate the influence of concentration, chain length, molecular weight distribution, chemical nature, solvent quality, aging effects, etc. on drag reduction as has been proposed, e.g., by Kulicke, Klein (1985).

6.3.3.1 Influence of the Stock Solution Concentration

In order to be able to make comparative studies between the series of measurements, one should always start from a stock solution of constant concentration. It has been shown that differing stock solution concentrations can also lead to significant differences in effectiveness (Fig. 20). Large discrepancies in the effectiveness characteristics arise at both concentrations 5 ppm and 15 ppm, depending on whether a 500 ppm or 1000 ppm stock solution has been used.

In order to obtain a higher effectiveness of drag reduction, it was therefore more convenient to produce the actual experimental solution from the more diluted stock solution. This is already an indication that the state of solution of macromolecules plays an important role in the interpretation of the results. However, a molecular weight of $10.1 \cdot 10^6$ g/mol and a concentration of 1000 ppm is just below the critical concentration c_1^* (this c_1^* is based on viscometric measurements) (Kulicke 1982). Above this critical concentration, mechanical entanglements ensure between the macromolecules. Beneath this limit, interactions between the molecules come into

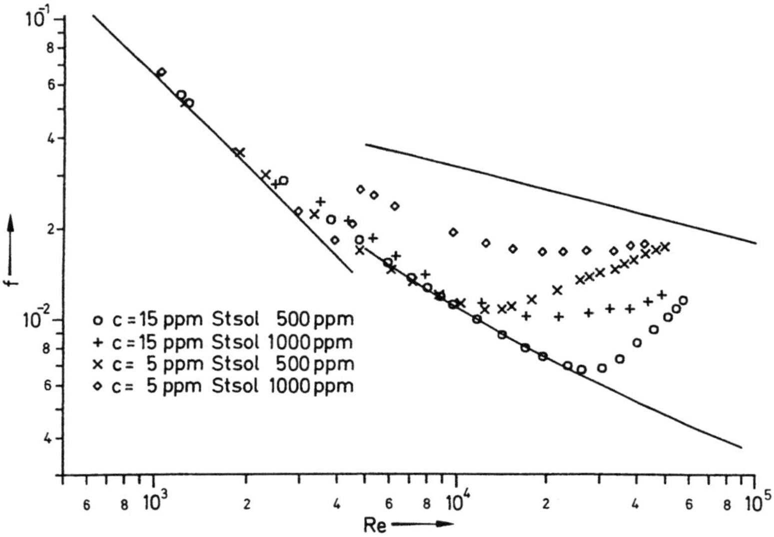

Fig. 20. Influence of the stock solution concentration (stsol) on the effectivity behavior of PAAm 37/05 in 0.1M Na_2SO_4 solution (M_w = 10.1 · 10^6 g/mol, T = 298 K, d = 2.08 mm)

effect which manifest themselves in the form of mutual interpenetration. Accordingly, the polymer-polymer-interaction parameter possesses a high value. The more dilute the solution, the smaller is the reciprocal influence between polymers, and the polymer-solvent interactions thus increase to the same extent, as observed by the positive effect on the flow enhancement. In principal, the solubility is improved, i.e., to attain the maximum effectiveness, each individual macromolecule must be completely dissolved.

6.3.3.2 Influence of Chain Length, Molecular Weight, and Molecular Weight Distribution

A high molecular weight is a necessary prerequisite for a good flow enhancer. This basic requirement is found as the result of many experiments. Yet a detailed investigation of the molecular weight dependency with clearly defined products has been carried out only by a few authors (e.g., Gramain; Gampert). The reason for this certainly stems from the fact that high-molecular-weight, commercial polymer products are very difficult to characterize. The application of sufficiently narrowly-distributed polymers is often not feasible, as considerable problems are likewise attached to the fractionation of higher molecular weights. To date, standards with a high molecular weight can only be used with polystyrenes. If specified reaction conditions are adhered to, polyacrylamides of the same distribution, but with differeing molecular weights, can be produced and chemically modified, so that a direct comparison of results is possible.

Figures 21 and 22 show, respectively, three and four different molecular weights and their effect as a dependant of the Reynolds number for two initial concentrations.

From these plots it becomes evident that it is important to measure the resistance characteristics over a large range of the Reynolds number, so as to be sure which section of the curve one is dealing with: at low Re-numbers is one still within the laminar-turbulent transition region? At high Re-numbers is one already near the maximum or already in the region of diminishing effectiveness? The measurements show that a higher effectiveness is obtained with increasing molecular weight.

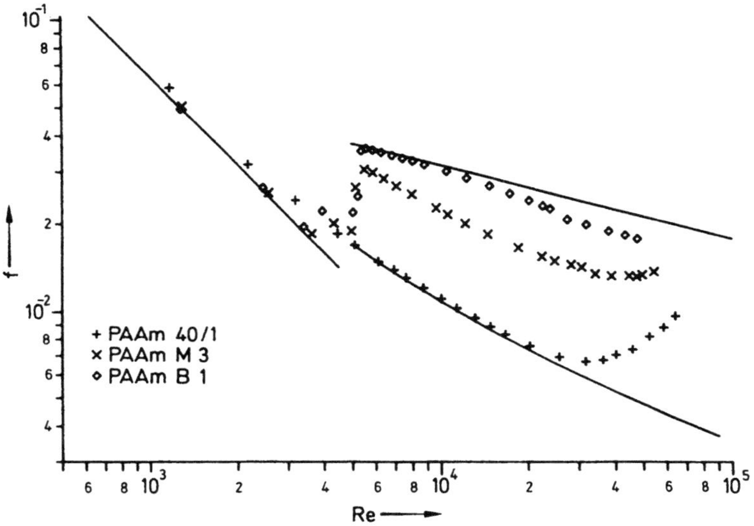

Fig. 21. Influence of the molecular weight at c = 30 ppm in distilled water; d = 2.08 mm PAAm B1 (M_w = 1.1 · 10^6 g/mol), M3 (M_w = 3.9 · 10^6 g/mol), 40/1 (M_w = 8.2 · 10^6 g/mol)

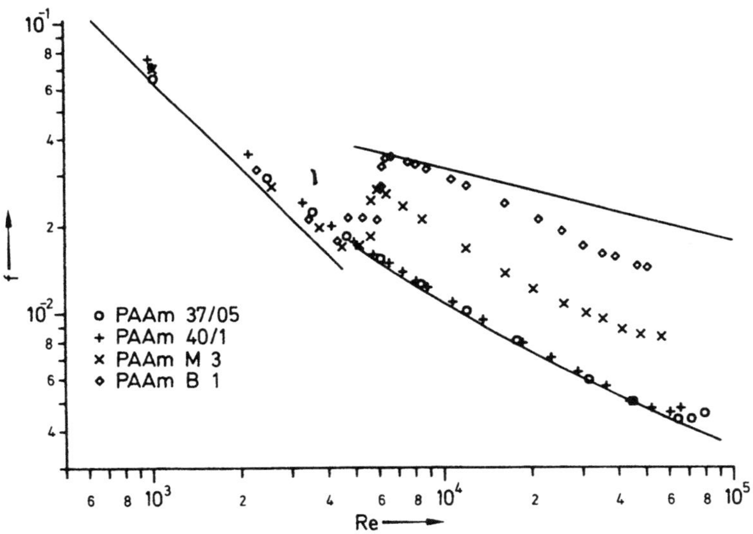

Fig. 22. Influence of the molecular weight, c = 98 ppm in 0.1M Na_2SO_4; PAAm 37/05 (M_w = 10.1 10^6 g/mol) (other M_w data see Fig. 21)

An interesting aspect arises in the transition region: for the polyacrylamides B1 and M3, definite changeover points are recognizable, whereas for polyacrylamide 40/1 there is a steady transition, so that a difference between laminar and turbulent flow cannot be pinpointed. For the curve of PAAm M3, the Newtonian turbulence reference-line is not attained. This phenomenon will be discussed in more detail in the following section on concentration dependency.

The abrupt change and onset behavior as well as the total trend of effectiveness during drag reduction is decisively influenced by the high-molecular-weight portion within a molecular weight distribution (Wagner). This is confirmed by investigations with blends of polyacrylamides, whereby the starting polyacrylamides B1, M3, and 40/1 are mixed in such a ratio that a M_w of $3.9 \cdot 10^6$ g/mol is obtained. Within the blend the proportion of high-molecular-weight PAAm 40/1 varies and it is precisely this proportion which determines the effectiveness: the larger this proportion is, the more effective the blend becomes and the greater the extend to which the turning point is displaced to lower Reynolds number. In the binary blend, the proportion of PAAm 40/1 is at its greatest and in the tertiary blend 3, at its lowest (see Fig. 23). As the initial concentration is always constant ($c = 30$ ppm), it follows that the effective concentration is determined by the high-molecular-weight portion within a distribution, i.e., the macromolecules with the longest chains are primarily responsible for the total course of the effectiveness.

Accordingly, the chain length represents the deciding factor for the extent of this flow phenomenon. When comparing the effectiveness of different chemically-structured polymers, it is therefore better to use the degree of polymerization P_w instead of the molecular weight. The former gives the number of linked monomer units and thus defines the chain length more distinctly. For example, polyacrylamide and polystyrene with a molecular weight of $6 \cdot 10^6$ g/mol possess different respective chain lengths and degrees of polymerization because of the unequal molecular weights of the monomers: for PAAm, $P_w = 84400$ (acrylamide = 71.08 g/mol) and for polystyrene, $P_w = 57600$ (styrene = 104.15 g/mol). The significant difference in chain length is a reason for the lower effectiveness of polystyrene under otherwise equal peripheral conditions. A precise statement concerning the effectiveness of a polymer is only possible by simultaneously specifying the degree of polymerization and distribution thereof, the Reynolds number, and the concentration.

In our case, the distributions of the degree of polymerization are equal and as the

Table 6. Effectiveness maxima for the PAAm's employed

Concentration: $c = 30$ ppm				
PAAm	B1	M3	40/1	37/05
P_w	15500	55000	115000	142000
M_w (g/mol)	$1.1 \cdot 10^6$	$3.9 \cdot 10^6$	$8.2 \cdot 10^6$	$10.1 \cdot 10^6$
DR_{max} (%)	17	40	72	73
RE	48000	25000	26000	26000
Concentration: $c = 98$ ppm				
DR_{max} (%)	32	60	77	78
Re	47000	41000	52000	64000

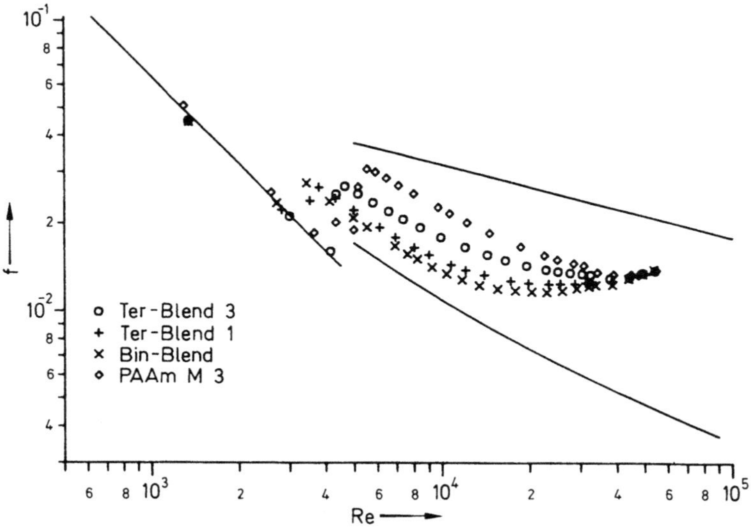

Fig. 23. Effectiveness behavior of PAAm blends. (c = 30 ppm in distilled water, $M_w = 3.9 \cdot 10^6$ g/mol, d = 2.08 mm). The part of the high molecular tail of the molecular weight distribution decreases from the left to the right side: Bind-Blend > Ter-Blend 1 > Ter-Blend 3 > PAAm M3

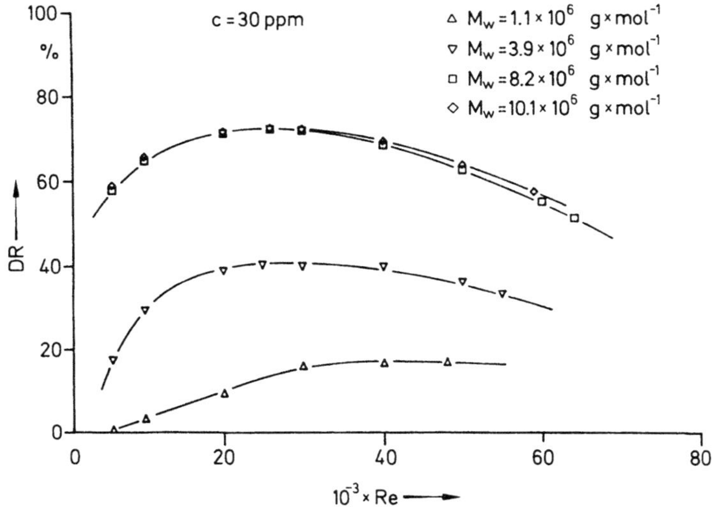

Fig. 24. Reduction in pressure loss as a function of the Reynolds number (d = 2.08 mm); polyacrylamide in distilled water

polymers employed were a homologous series of polyacrylamides, a higher molecular weight is therefore synonymous with a larger chain length.

The maximum attainable reductions in drag resistance and their corresponding peripheral conditions are summarized in Table 6. The list of values shows that the respective effectiveness maxima of individual molecular weights lie at different Reynolds numbers (see Fig. 24).

6.3.3.3 Dependence on Concentration

The higher the initial concentration, the greater is the effectiveness of the polymer and, by analogy, the maximum of effectiveness is displaced in the direction of increasing Reynolds numbers. At equal concentrations, the long-chain macromolecule is more effective, as in fact one would expect for reasons of the chain-length influence.

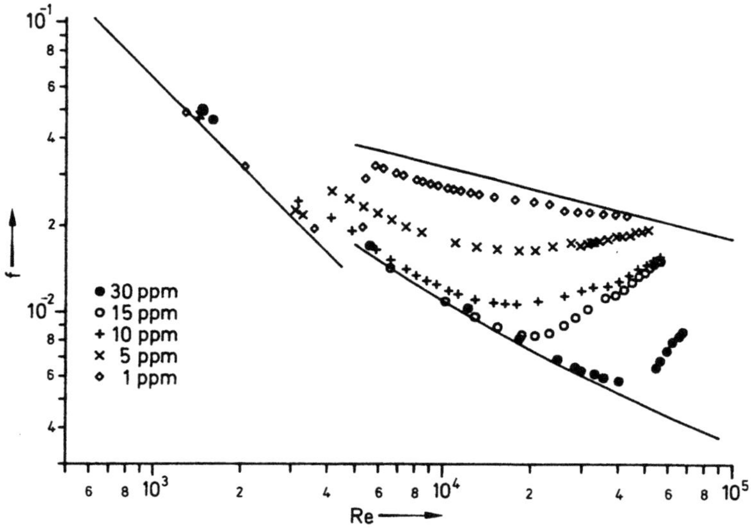

Fig. 25. Influence of polymer concentration on reduction in pressure loss for PAAm 40/1 (M_w = 8.2 · 10^6 g/mol, T = 298 K, d = 2.08 mm) in 0.1*M* Na_2SO_4 solution

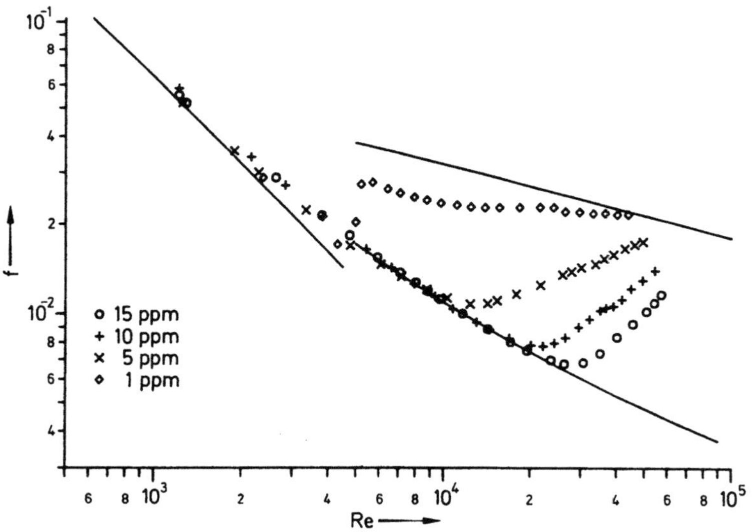

Fig. 26. Influence of polymer concentration on reduction in pressure loss for PAAm 37/05 (M_w = 10 · 10^6 g/mol, T = 298 K, d = 2.08 mm) in 0.1*M* Na_2SO_4 solution

After passing through the maximum, the curves bend off towards the turbulence reference line for Newtonian fluids. This increase in the friction factor observed at high Re numbers can be traced back to the influence of pipe roughness (Bewersdorff, Berman 1987). In the case of smaller polymer starting weights, this is almost reattained. At the very low concentration of 1 ppm, an stonishing pressure drop of 15% (Re = 9500) is caused by PAAm 40/1 and even 27% (Re = 8000) by PAAm 37/05. This is an impressive confirmation of the extraordinarily high effectiveness of polymeric flow enhancers.

Very interesting changes in flow behavior arise when comparing Newtonian fluids and polymer solutions in the laminar-turbulent transition area. At the concentration of 1 ppm, the change-over or onset point even occurs at the same Re number as for Newtonian fluids, though the measuring points already lie below the turbulence line. In the case of PAAm 40/1, at 5 ppm the change-over point is displaced towards lower Re numbers. For higher concentrations, the onset point is no longer recognizable, the transition from laminar to rubulent flow crosses over continually without a take-off point occurring, as was also found by several authors (Interthal 1984, Bewersdorff, Singh 1987; Ogawa). In this case the critical wall shear stress is exceeded in the laminar flow region.

This phenomenon makes it possible to differentiate between solutions according to their effective behavior. If the change-over point is still recognizable as a distinct take-off point, then we can speak of "diluted drag-reducing" solutions, whereas, if a constant transition is the case, then "concentrated drag-reducing" solutions are concerned (Zakin). In this latter case, it cannot be stated at which Reynolds number turbulent flow begins and where laminar flow ends.

In the discussion of the mechanism of drag reduction, one question that commonly arises is whether this phenomenon is brought about by individual molecules or aggregates (Dunlop; James 1980). To define a boundary between the respective ranges for dilute and moderately concentrated solutions, it is useful to specify a critical concentration c*. The precise specification of molecular weights, limiting viscosity values, and other molecular parameters requires that the measurements be performed in dilute solutions (particle solutions). Intermolecular interactions in the moderate concentration range can cause considerable discrepancies within the recorded data. Several methods exist for calculating c*, depending on which model is taken as a basis for describing the macromolecule. Some definitions are presented in the next section.

If Einstein's law (8) for spherical particles is taken as the starting point, one obtains a critical concentration as presented in Eqs. (8) and (9).

$$(c \cdot [\eta])^* = 2.5 \tag{8}$$

$$c_1^* = \frac{2.5}{[\eta]} \tag{9}$$

However, Eq. (8) leads to the paradox that overlapping does not begin until a critical volume fraction of 2.5. In order to overcome this, somewhat modified equations have been put forward by Simha (10)

$$c_2^* = \frac{1}{[\eta]} \tag{10}$$

and Onogi and co-workers (11)

$$c_3^* = \frac{0.7}{[\eta]} \tag{11}$$

by application of a lattice model.

More recent investigations show that these descriptions are oversimplified and lead, therefore, to incorrect results, i.e., it is not correct to calculate the critical concentration from the volume of the molecules at an intrinsic concentration. Moreover, with increasing concentration, polymer-polymer interactions become decisive and lead in the case of a thermodynamically good solvent to coil shrinkage and hindered interpenetrability. Therefore, the range of ideally dilute solution is actually smaller than approximated with Eqs. (9) to (11). In these cases one obtains semi-dilute interacting polymer solutions, as predicted by DeGennes, of which the rheological behavior is strongly influenced by excluded volume effects. This has been shown experimentally by various other authors (e.g., Boog; Kulicke, Kehler, Bouldin).

A satisfactory description of the state of solution is therefore only obtained in the case of a thermodynamically good solvent by taking five distinct states of solution into consideration, namely ideally-dilute particle solutions, semi-dilute particle solutions, semi-dilute network solution, concentrated particle solution, and concentrated network solution (Fig. 27).

Analysis of the viscous and elastic behavior of polymer solutions show that the macromolecules experience intermolecular interactions at concentrations that are

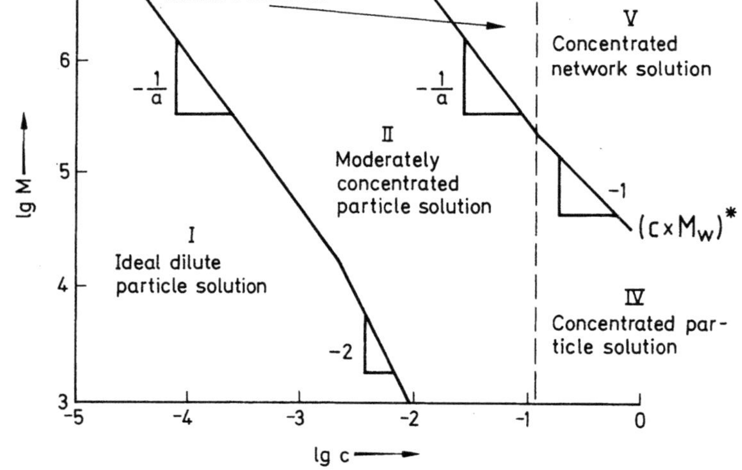

Fig. 27. States of solution as a function of the polymer concentration and molecular weight for a thermodynamically good solvent

much smaller than those previously assumed (Kulicke 1986). Analysis of the separate terms of the η_0-M_w-c relationship (Kulicke, Kniewske 1986) gives rise to the following critical concentration (Bouldin), i.e., intermolecular interactions become decisive at an overlap volume much smaller than 1.

$$c_4^* = \frac{10^{-2}}{K_H \cdot [\eta]} \qquad (12)$$

Furthermore, the onset of polymer solvent interactions is solvent-dependent due to the fact that K_H is a function of the solvent quality (Gundert).

The critical concentration c* (Bouldin) denoting the onset of intermolecular entanglements and c** (Graessley), which describes the transition from a semi-dilute to a concentrated solution, are of no interest in the field under consideration due to the fact that the applied concentrations are far smaller in every case.

In the discussion of critical concentrations it is important to know which molecular model is being used and which type of measurements are to be carried out. In general, the series arranged according to size is as follows: $c_1^* \gg c_2^*, c_3^*, c_4^*$. For the case of polyacrylamides in particular the following series is valid (Kulicke 1982):

$$c_1^* \gg c_2^* > c_3^* > c_4^*$$

When performing viscometric measurements, the initial concentrations are chosen so as to be less than or equal to c_1^*, for light scattering investigations they lie below c_1^* (cf. Fig. 14). It clearly emerges, from a comparison of the defining equations, that all concentrations used for flow measurements lie significantly below these critical

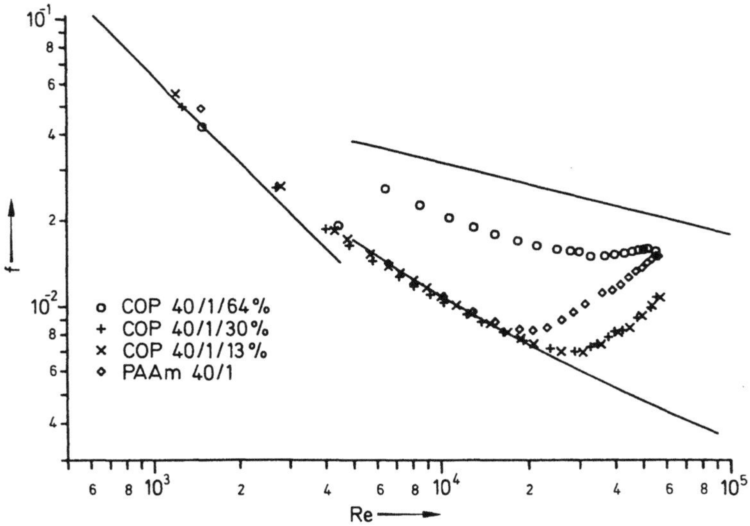

Fig. 28. Influence of degree of saponification for the homologous series 40/1, (c = 15 ppm in 0.1*M* Na_2SO_4 solution, d = 2.08 mm)

values. Accordingly, experiments are carried out with highly dilute solutions thus justifying the assumption that drag reduction is caused by individual molecules.

6.3.3.4 Influence of Chemical Nature

In order to be able to record the influence of the chemical nature on the drag reduction not only qualitatively, but also quantitatively, the molecular weight distribution must be known in addition to the absolute molecular weights.

This is as a rule not possible as these data are not available for high-molecular-weight samples. In order to be able to exclude the influence of the distribution of degree of polymerization for various polymer structures, polymer analogous reactions may be carried out, where polymers of differing chemical structures and molecular weights, but with the same distribution of degree of polymerization are obtained (e.g., Kulicke, Hörl 1985). Of particular interest in this report are polyelectrolytes.

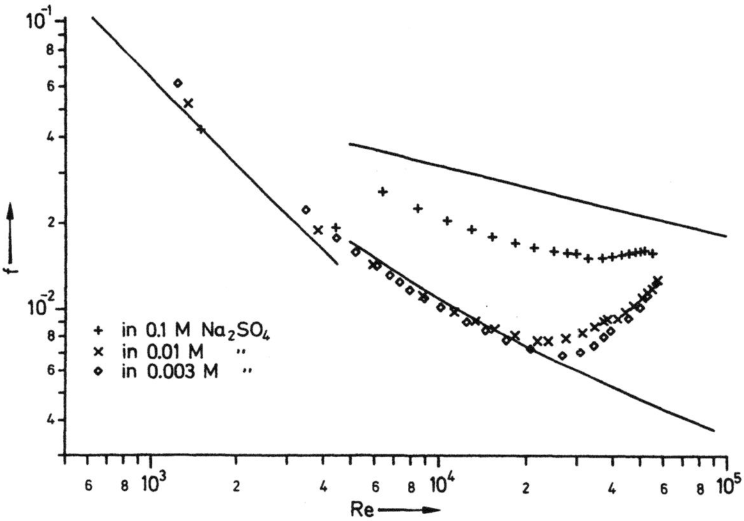

Fig. 29. Influence of electrolyte salt content on the effectiveness of coacrylate 40/1/64%, (c = 15 ppm, d = 2.08 mm)

The aqueous solutions of polyelectrolytes represent an energy-rich and very sensitive system. This has become apparent once already in the previously discussed aging process as well as addition of low-molecular-weight additives, such as salts, or during a change in pH value. The macroscopically observable changes in viscosity behavior which occur in these instances point to significant restructuring of solution behavior. The effects of this restructuring become more drastic the greater the degree of saponification and hence the quantity of charge carriers becomes. To obtain reproducible results therefore, the peripheral conditions must be precisely observed. Figures 28 and 29 depict the influence of the degree of saponification on the effectiveness of the polyacrylamide-coacrylates in the presence of high electrolyte concentrations (0.1*M*

Na_2SO_4). It can be clearly seen that within the homologous series 40/1, the effectiveness of the low-saponified samples is greater than that of the starting polyacrylamide. A difference between 13% and 30% saponified coacrylates is, however, barely distinguishable. A differentiation in the direction of decreasing effectiveness is possible solely for c = 5 ppm:

$$\text{COP } 13\% > \text{COP } 30\% > \text{PAAm } 40/1 > \text{COP } 64\% .$$

The coacrylate 40/1/64% is less effective under these experimental conditions, despite its larger coil volume (higher Staudinger index) than the starting polyacrylamide 40/1, a feature which can be seen particularly clearly in Fig. 29 at c = 15 ppm. The maximum effectiveness lies beneath the given requirements in the range of lower saponification levels. Accordingly, correlation with maximum behavior, as found in viscometry, is not possible. The curves of Fig. 29 unanimously show that the drag-reduction effect of polyelectrolytes increases with reducing salt content. However, an increase in hydrodynamic volume is also connected with this. The solution viscosity of the coacrylate is significantly greater in aqueous medium than in the presence of salts.

The loss in effectiveness in 0.1*M* Na_2SO_4 is particularly drastic. The high salt concentrations must surely influence the solvation characteristics of the polyelectrolyte and change the thermodynamic quality of the solvent. The effects on the solution structure and the flexibility of the polymer backbone must also be discussed in this connection, as, for example, suggested by Brostow (1984).

According to his interpretation, both the degradation characteristics and the extent of drag reduction are determined by the relationship between compact and flexible chain segments. As a consequence of this, it is not the absolute volume of the (gel) coil which is the decisive value, but the behavior of the polymer chain.

In each case, the highest effectiveness of polyelectrolytes is obtained in distilled water and at low salt contents.

Due to the measurements obtained, the following interpretations for the mechanism of polyelectrolytes are supported. By introducing ionic groups into the backbone of the polymer, the (gel) coil volume of the macromolecule increases and the degree of coiling decreases. The molecule can, as a consequence of this, be subjected to the shear and elongational stresses arising in turbulent flow at an earlier stage. This is because it also possesses a larger working surface due to the greater hydrodynamically effective volume.

The coacrylates can thus be deformed more easily and aligned in the direction of flow.

The end effect is that a higher degree of elongation is obtained due to the greater capacity for deformation and to the greater, in comparison to the polyacrylamides, relaxation times (see Sect. 4.7). This degree of elongation in turn leads to a longer, hydrodynamically effective chain which again causes increased effectiveness.

As Hashemzadeh's measurements in porous media flows have shown, the coacrylates have a smaller elongational viscosity portion than the polyacrylamides, i.e., they are more easily deformed and their drag-increasing mechanism first comes into effect at smaller Reynolds numbers, although the value of the drag increase at this point is far smaller than for the polyacrylamides.

In contrast to the copolymers, the drag-reduction efficiency of polyacrylamides is in no way influenced by differing high electrolyte concentration, as is confirmed by the measurements for PAAm 37/05 at c = 5 ppm and c = 15 ppm (Fig. 30).

When using bivalent salt cations, in this case magnesium sulfate ($MgSO_4$), the effectiveness of the coacrylate 37/05/64% decreases with increasing salt concentration. The effectiveness in the presence of the bivalent cation is already distinct when comparing the attainable drag reductions in 0.002*M* sodium sulfate and 0.001*M* magnesium sulfate solutions (s. Fig. 31).

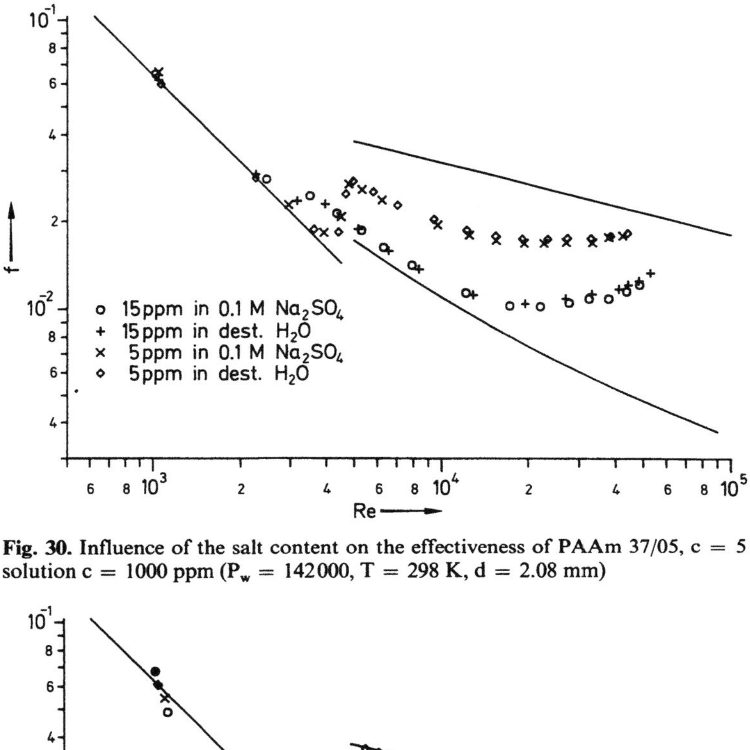

Fig. 30. Influence of the salt content on the effectiveness of PAAm 37/05, c = 5 and 15 ppm, stock solution c = 1000 ppm (P_w = 142000, T = 298 K, d = 2.08 mm)

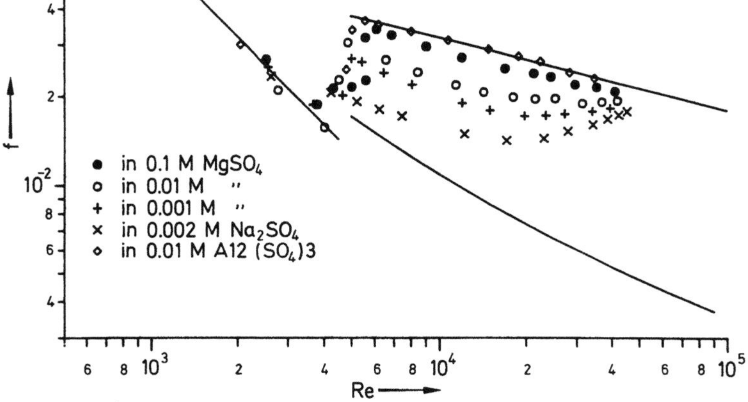

Fig. 31. Influence of polyvalent salt cations on the effectiveness of the coacrylate 37/05/64%, c = 5 ppm (P_w = 142000, T = 298 K, d = 2.08 mm)

The flow enhancement effect is completely suppressed in the presence of trivalent cations. Even at the preparation stage for aluminium sulfate ($Al_2(SO_4)_3$), the polymer is precipitated at a concentration of c = 5 ppm: the flakes thus formed can be clearly seen, suspended in solution. Consequently, no drag-reduction effect can be expected and the experimental solution behaves as a Newtonian fluid (Fig. 31). Therefore, it is likely that multivalent cations from complexes or ionotropic gels and that by exceeding a critical value they may precipitate out of solution.

Other water-soluble copolymers were recently strongly investigated with regard to their drag-reduction behavior. McCormick synthesized some well characterized model acrylamide copolymers. He found that all copolymers conforms to a universal curve when normalized for volume fraction of polymer in solution. Bock observed that the drag reduction is decreased by adding 10–1000 ppm diblock or triblock tertiary-butyl-styrene-sulfostyrene polymer. Experimental studies with polyelectrolytes were undertaken by Schulz. Here, it was confirmed that a drag reduction occurs if 0.009 %–0.1 % of the copolymer were added with a mixture of 90–99 mol-% acrylamide and 1–60 mol-% of the following ammonium compound:

$$H_2C = C(CH_3)CO_2CH_2CH_2N^+(CH_3)_2(CH_2)_3SO_3^-$$

The investigations by Peiffer relate to improved drag-reduction agents for aqueous solutions which comprise a polymeric complex which is a terpolymer of acrylamide/metal styrene sulfonate/methylacrylamidepropyltrimethyl ammonium chloride (MAPTAC). The metal styrene sulfonate is an anionic monomer while MAPTAC is cationically charged. These acrylamide-based polyampholytes have approximately 1 to about 50 mol-% of the anionic monomer and approximately 1 to about 50 mol-% of the anionic monomer and approximately 1 to about 50 mol-% of the cationic moiety present within the macromolecular structure. These groups are not necessarily present in an equimolar charge ratio. The excess undissociated charge allows for facile dispensability of solubility of the polyampholytes into fresh water.

The results received so far support the view that the state of solutions of the individual macromolecules must be discussed very exactly with regard to this polymer-induced turbulent flow phenomenon. Among the essential factors which must be taken into consideration are the chemical nature of the polymeric flow enhancer as well as all parameters which affect the solvation characteristics and chain mobility.

In contrast to this, Armstrong reported that for dilute solutions (≦10 ppm), the drag reduction is independent of the chemical nature of the polymer and relates only to effects on solution viscosity.

6.3.4 Degradation

The presence of small amounts of certain polymers can produce spectacular reduction in the frictional losses of fluids in turbulent flow through conduits. Drag reduction has an immense field of applications, both currently and potentially. The list of exploitable situations as described in Sect. 2 could be extended, but a big snag exists: drag reduction decreases with flow time. This is believed to be due to mechanical degradation of added polymer (Brostow 1983). In Fig. 32 and Fig. 33 the influence of M_w on drag reduction is displayed.

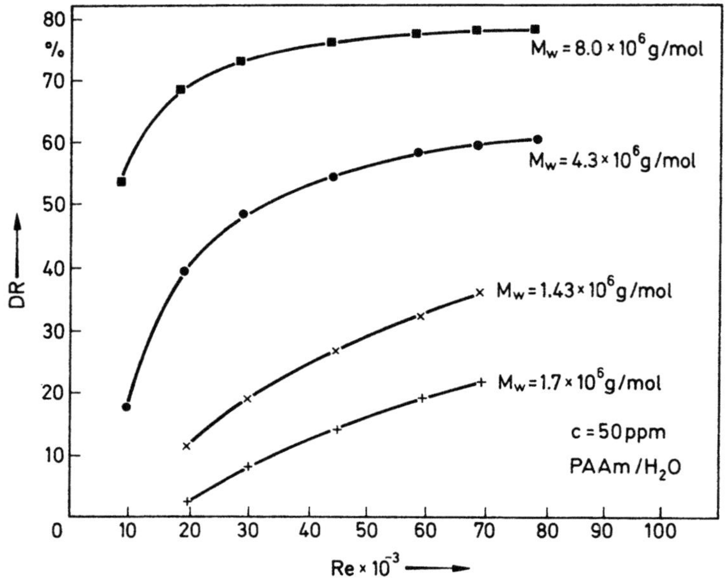

Fig. 32. Influence of M_w on drag reduction effect for nonionic polyacrylamide samples, d = 6.1 mm (data taken from Gampert, Wagner 1985)

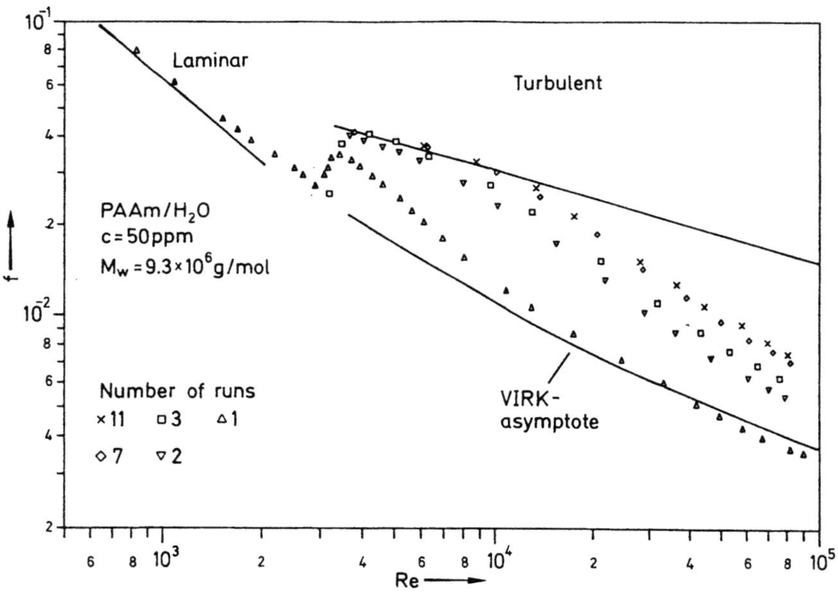

Fig. 33. Influence of the number of runs on the effectivity of PAAm in water, d = 6.1 mm (data taken from Wagner)

It seems, therefore, useful to take the numerically simple laminar shear flow as a starting point for the description. As recently published (Bouldin), an easily manageable model has been derived which enables an *ad hoc* prediction to be made of the critical shear rate at which mechanical chain scission takes place (cf. Fig. 34).

As Bouldin has pointed out, degradation occurs when all intermolecular coupling points are loosened and the discrete molecule is subjected to the velocity field as shown in Fig. 35.

This critical value can be estimated by application of the η_0-M_w-c relationship, due to the fact that the viscosity level of a completely disentangled polymer solution is equal to the product $c \cdot [\eta]$, i.e., the specific viscosity is directly proportional to

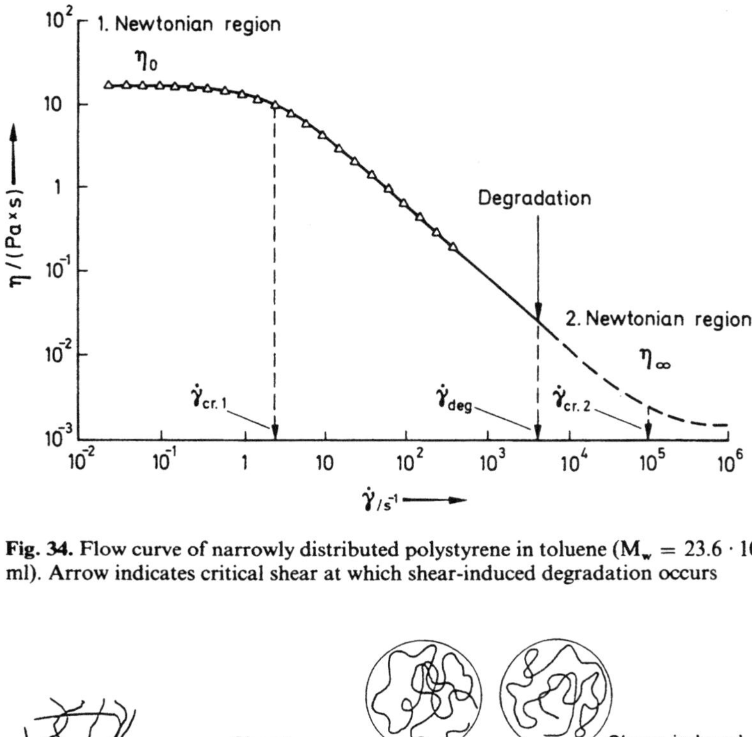

Fig. 34. Flow curve of narrowly distributed polystyrene in toluene ($M_w = 23.6 \cdot 10^6$ g/mol, c = 0.1 g/ml). Arrow indicates critical shear at which shear-induced degradation occurs

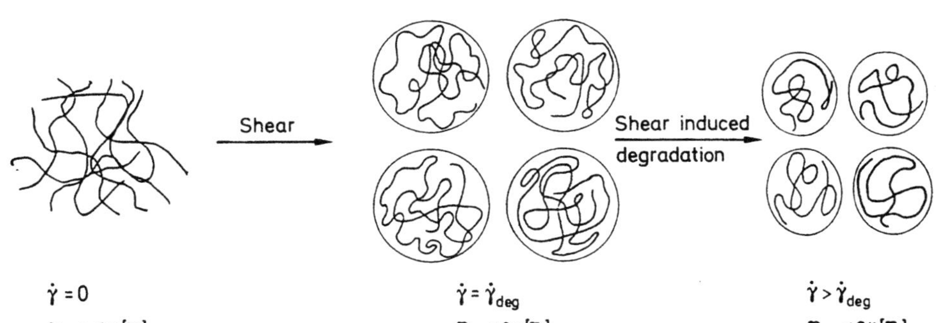

Fig. 35. Schematic picture of the state of solution of a non-sheared solution, a disentangled solution at the critical shear rate and a degraded sample

the effective volume fraction polymer. Further enhancement of the shear gradient leads to a reduction in the mean molecular weight. The latter can easily be determined with the above-mentioned statement.

$$\eta_{sp}(\dot{\gamma}) = c \cdot [\eta]_{deg} \tag{13}$$

$$M_{deg} = \left[\frac{\eta_{sp}(\dot{\gamma})}{c \cdot K_\eta}\right]^{1/a} \tag{14}$$

Obviously, this approach does not only take into consideration the influence of the molecular weight and concentration, but also the contribution of the solvent power on the state of solution. With these shear stability criteria, it is therefore possible to make *ad hoc* predictions on the basis of the data of the single chain, which can be taken from handbooks in many cases. In addition, it should be noted that theoretical forecasts fit the experimental data very satisfactorily (Kulicke, Kniewske 1986).

To describe the mechanical degradation in turbulent flow, two limiting cases must be introduced: the laminar shear flow and the elongational flow.

In order to estimate to what extent degradation actually occurs in a turbulent pipe flow, we carried out measurements at different Reynolds numbers, as shown in Fig. 36.

The results imply that in a transition region from laminar to turbulent flow, the degree of degradation seems to decrease. However, it should be noted that the effect is relatively small and there is still some doubt as to what extent the data are reproducible.

Works published by Brostow (1984) as well as by Kim (1986) seem to indicate that the chemical structure correlates with the stability of a macromolecule in turbulent flow. As mentioned above, the drag reduction of polymeric fluids seems to be linked

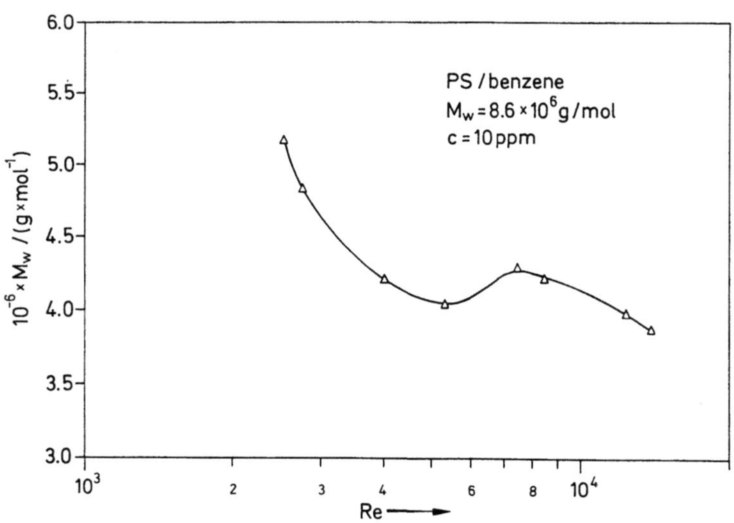

Fig. 36. Mean molecular weight of a polymer after flow through a pipe as a function of the Reynolds number (d = 2.08 mm)

to the extensibility of polymer coils. Assuming the elongational viscosity to be the determining factor for drag reduction, Durst, Haas (1982) have proposed that the elongation rate correlates with the Reynolds number of the fluid and that, therefore, the effectiveness should be linked with the Deborah number De. The latter is defined as the ratio of the relaxation time of the macromolecule to the time between two elongations $(1/\dot{\varepsilon})$.

$$\mathrm{De} = \tau \cdot \dot{\varepsilon} \tag{15}$$

Starting from the FENE model (Warner; Bird and collaborators) one obtains the following statements:
(i) the relaxation time depends on the concentration, molecular weight, and solvent quality;
(ii) the model predicts a finite extensional viscosity. The fundamental soundness of

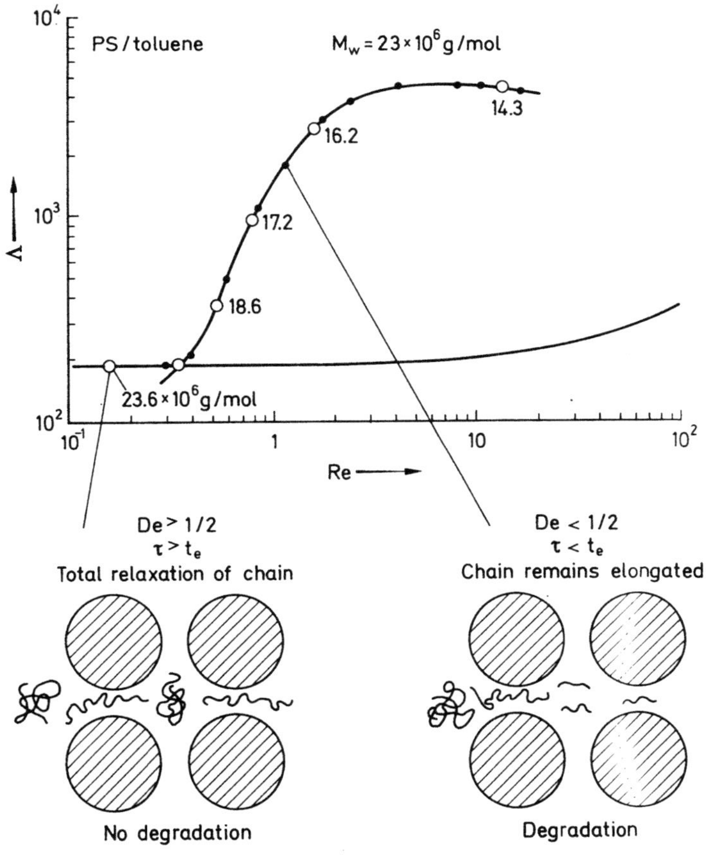

Fig. 37. Experimental course of the resistance coefficient f as a function of the Re number for narrowly distributed polystyrene ($M_w = 23.6 \cdot 10^6$ g/mol, c = 100 ppm) in toluene and the mean molecular weight after passing through the porous medium

this approach in describing elongational flow situations has been proven experimentally by Kulicke, Haas (1984) by producing elongational flow in porous media. However, it has been proposed by Durst (personal communication) that this non-Newtonian flow should be accompanied by elongationally induced degradation. He further argues that presumably this fact leads to the decrease of the resistance factor beyond the maximum point. Investigations carried out by Hashemzadeh reveal, however, as shown in Fig. 37, that the onset of elongational flow is directly connected with mechanical degradation of the macromolecules.

Furthermore, it is worth remarking that Kulicke, Haas (1984) have found that the critical Reynolds number, which defines the onset of tensional stress, is strongly shifted in the case of PAAm if the measurement is repeated (2nd run), i.e., the molecule is degraded, whereas in the case of partially saponified PAAm/AAcNa, no pronounced change of Re_0 is to be found, as shown in Fig. 38. This phenomenon has been attributed to the relatively high rigidity of the partially hydrolized polyacrylamide.

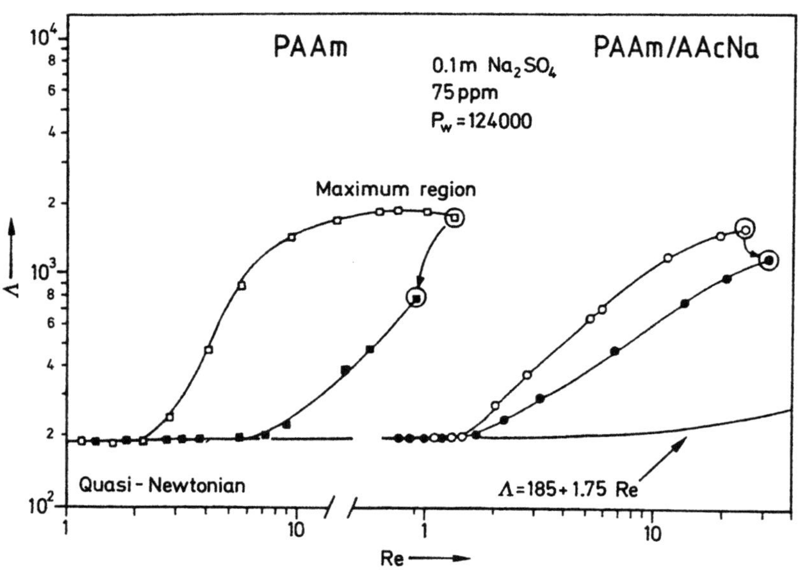

Fig. 38. Friction coefficient of PAAm and partially saponified PAAm/AAcNa as a function of the Re number for the first and second run (taken from Kulicke, Haas 1984)

In conclusion, it can be stated that in both laminar and turbulent flow, polymer elongation can become the determining flow feature. As shown in Fig. 39, an extension of the molecule in laminar flow leads to a drastic increase in the resistance coefficient. This effect is accompanied, as mentioned above, by chain scission.

In turbulent flow, the macromolecules are also believed to experience burst-induced elongation. However, it has to date not been possible to formulate structure-property relationships which allow an *ad hoc* prediction of the onset of mechanical degradation and its influence on drag reduction.

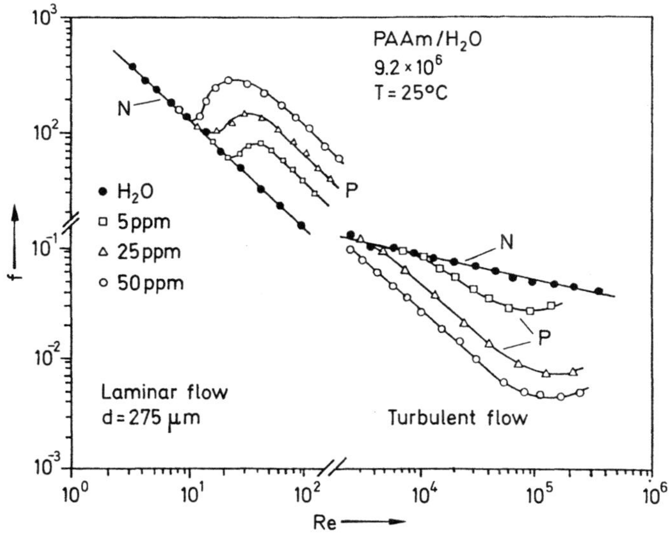

Fig. 39. Course of the resistance coefficient in laminar and turbulent flow as a function of the Reynold number

7 Conclusions

The aim of the work under consideration was to correlate the effectiveness of the drag reduction with the molecular parameters of the polymer causing it.

As it has not so far been possible to predict this phenomenon in a clear and quantitative manner, its technological use has yet to be fully realized. As far as is known, the only large-scale technical application is the Alaska-pipeline (see Sect. 2).

However, as was shown at the Iutam Symposium in Essen in 1984 and the subsequent 3rd International Conference on Drag Reduction in Bristol in 1984, a wide variety of large-scale technical applications are offered. Of these many possibilities, let it suffice to mention here the application of polymers in sewage systems (e.g., Dembek) and the hydro-transportation of coal in pipelines (e.g., Golda).

A further potential field of application may lie in the area of refrigeration and heating technology (e.g., Kemmer; Durst 1985), but it has not been possible to enter into details of this in the present work. As the problems are manyfold, it seems sensible to treat this area as an independent study.

A survey of the models presented to date (see Sect. 4) shows that it has not yet been possible to make a quantitative statement concerning the influence of molecular parameters, neither has it been possible to put forward a structure- property relationship.

The experimental techniques were dealt with briefly in Sect. 5, where it must be noted that, for both the quantitative and qualitative description of the drag-reduction phenomenon, precisely defined measuring equipment must be used in order to be able to establish whether degradation arises as a result of the nature of the flow of the apparatus as such (corners, edges, burrs, surface, roughness).

The main interest of this work, therefore, lay in the field of water-soluble polymers and their influence on drag reduction in turbulent pipe flow, as it is from this field that the major technological use is to be expected. The investigation of the influential parameters for this class of polymers, such as molecular weight and distribution thereof, thermodynamic quality of the solvent, to name but a few (see Sect. 6.3.3), must precede a clear-cut characterization (see Sect. 6.3.1) of the polymer used.

This is made particularly difficult in the case of water-soluble polymers for reasons of their chemical nature. The phenomenon of polymer aging should also be mentioned in this connection (see Sect. 6.3.2).

The experiments with highly dilute solutions have shown quite clearly that individual macromolecules are capable of causing drag reduction. In this case it was found that a PAAm solution even at a concentration of 1 ppm ($M_w = 8 \cdot 10^6$ g/mol) can lead to a 15% drag reduction. The chain length is primarily responsible for the extent of this turbulence phenomenon. A loss in chain length cannot be compensated for by a rise in initial concentration in order to obtain equal effectiveness values. A knowledge of the degree of polymerization, which permits an exact statement of the chain length, is a necessary though not sufficient prerequisite.

The effectiveness of a polymeric flow enhancer is influenced decisively by the state of solution and the solvation characteristics. In the case of polyelectrolytes, in particular, the chemical nature plays a significant role, e.g., it was found for poly(acrylamide)-coacrylate that a significant increase in effectiveness arises with the increasing number of ionic groups. It is therefore necessary to consider, for example, such factors as: the question of critical concentration, polymer-polymer and polymer-solvent interactions, the thermodynamic quality of the solvent, the proportion of ionic molecular groups and their behavior in the presence of lower-molecular-weight charge carriers.

Basically all factors are important which display an effect on the solubility of the macromolecule and its flexibility, i.e., the capacity for deformation of the polymer backbone. According to Brostow's theory, there are two effects of central importance for practical application: degradation and relaxation. Every type of energy transferred to a polymer chain in solution does so via one of these two processes. Thus, all parameters which directly affect these processes are of corresponding significance.

Subsequent to the evaluation of flow measurements, the following assertions can be made about polymer behavior in turbulent flow.

The determining criterion for drag reduction, similar to fibers, is a large length-diameter ratio. The realization of this, hydrodynamically speaking, anisotropic shape requires that a macromolecule which exists as a coil at rest be elongated to its farthest possible extent. For the polymer molecule to be affected or extended at all in turbulent flow, the polymer relaxation times and the charactistic flow times must bear a specific relationship to one another. The coil can be fixed between the eddies and elongated in the direction of flow. The relaxation time should therefore be greater or at least equal to the life time of the eddy. However, a specified molecule size is also coupled to this. If the polymer time is smaller, the macromolecule cannot be stressed because it will always return to its starting condition in the stress phase.

The shape of the molecule should differ from that of a sphere in this phase, as a sphere may always evade the reacting forces by rotation. Deformation produces a directional bias which enables the molecule to be aligned and fixed in the direction of flow and thus allows elongation to take place. As the coil shape must deviate from a spherical geometry in order to act as a drag-reducing agent, the specified chain length and hence molecular weight must exceed $M_w = 1 \cdot 10^6$ g/mol, as has already been confirmed by numerous measurements. On the other hand, another absolute requirement for reciprocal influence is that the minimum chain length lie within the order of magnitude of the eddy dimensions.

The basic properties for a good polymeric flow enhancer accordingly read as follows:
It must display a high degree of polymerization and a flexible backbone, as well as good solubility in the appropriate solvent, one which leads to the greatest possible coil extension. This trend can be improved by the introduction of ionic molecular groups with the same structural orientation of charges. Because of the rise in (gel) coil volume, the molecule is activated in turbulent flow at an earlier stage and can be more easily deformed and aligned in the direction of flow. Linked to this is stretching of the molecule, which leads to a thread-like shape. The polymer thread subdues the momentum exchange and suppresses the growth and diffusion of eddies.

This accounts for the substantially higher effectiveness compared to the fibers: as a consequence of their microscopic dimensions, the macromolecules are able to engage in the production and growth stages of eddies considerably earlier, thus exerting a sustained damping influence on the axial and radial turbulence structure, as conformed by *LDA* measurements in injection experiments.

If the duration and intensity of the elongational stress are sufficiently large, the macromolecule can be stretched to a greater extent. Keller was able to prove this using a cross-slot device producing a planar elongational flow field with the aid of flow diffraction.

Dimensional analysis shows that the measurements of the smallest eddy L_K and the size of the strained molecule L_{max} lie within the same order of magnitude.

The following statement is taken as a basis for assessing the length of macromolecules.

The maximum attainable chain length L_{max} (Eq. 18) can then be calculated via the degree of polymerization.

$$P_w = \frac{M_w}{m_0} \tag{16}$$

$$L_{max} = l_{mon} \cdot P_w \tag{17}$$

With the aid of the Kolmogorov micro-measures for eddy size L_k and the eddy life time t_K, the following relationships, Eqs. (19) and (20) follow:

$$L_k/d = (xy^+)^{0.25} \cdot 8^{0.5} \cdot Re^{-1} \cdot f^{-0.5} \tag{18}$$

$$t_K \cdot \gamma/d^2 = 8 \cdot (xy^+)^{0.5} \cdot Re^{-2} \cdot f^{-1} \tag{19}$$

The Kolmogorov relationships are not valid in immediate proximity of the wall, as the turbulence is strongly anisotropic in this region. For a first assessment of the order of size of the turbulence measurements, the point of maximum dissipation is chosen.

Hlavacek arrived at similar deductions using the Virk length-mass theory as a basis and refining it. Virk and his collaborators worked on the assumption that the eddies were bigger than the macromolecules by a factor of about 1000. The diameter of the eddy at that time was in the region of $1 \cdot 10^{-2}$ mm. Proceeding from the study of eddy production, Hlavacek postulated measurements for the eddy precursor of $1 \cdot 10^{-5}$ mm. The radii of gyration for most of the polymers (see Sect. 6.3.1) already lie within this region. According to the interpretations of the above-mentioned authors, the active eddy precursors are suppressed by strong associations between solvent molecules and polymer chains. Consequently, the extent of drag reduction is considerably influenced by polymer-solvent interactions, whilst also taking into account the molecular weight concentration and flow geometry. On the other hand, a definite requirement for a reciprocal influence is a minimum chain length, which should lie within the order of size of the eddy dimensions.

Due to the long relaxation times of the macromolecules in comparison to the turbulence times, a deformed macromolecule can readily outlive the lifespan of a micro-eddy and thus exert a decisive influence on the latter. On the other hand, this also means that extension of the molecule will be made possible by the combined effects of many consecutive generations of eddies. A large proportion of the turbulent energy is thereby used in the alignment and deformation of the polymer molecule.

However, the long range effectiveness of polymer additives remains, due to the mechanical degradation, a hitherto unsolved problem. By application of the above-mentioned theoretical approaches and the influence of laminar and elongational flow on polymer stability described in Sect. 6.3.4, it seems possible to retain the flow features over a longer period. It is therefore necessary to reinforce investigations which enable a more quantitative description of turbulent flow, so that in the future structure-property relationships can be established which permit a correlation of the microscopic structure of the macromolecules with the observed flow phenomena.

The authors thank Prof. F. Durst and Dr. H. W. Bewersdorff for their valuable comments and helpful discussions.

The support of this work by the Deutsche Forschungsgemeinschaft (DFG) and by the Fonds der Chemischen Industrie (FC) is gratefully acknowledged.

8 References

1. Armstrong R, Jhon MS (1984) "A self-consistent theoretical approach to polymer induced turbulent drag reduction" Chem Eng Commun 30: 99
2. Astarita G (1965) "Possible interpretations of the mechanism of drag reduction in viscoelastic liquids" Ind Eng Chem Fundam 4: 354
3. Balakrishnan C, Gordon RJ (1975) "Influence of molecular conformation and intermolecular interactions on turbulent drag reduction" J Appl Polym Sci, 19: 909
4. Banijamali SH, Merrill EW, Smith KA, Peebles LH (1974) "Turbulent drag reduction by polyacrylic acid" AIChE J 20: 824
5. Beatty WR, Wheeler GR, Johnstone RL, Kramer RL, Warnock LG (1984) "Offshore crude oil production increased by drag reducers" 3rd Intern Conf on Drag Reduction, paper F1, Bristol, U.K.
6. Berman NS, George WK (1974) "Onset of drag reduction in dilute polymer solutions" Phys Fluids 17: 250
7. Berman NS (1977) "Flow time scales and drag reduction" Phys Fluids 20: 168

8. Berman NS (1978) "Drag reduction by polymers" Ann Rev Fluid Mech 10: 47
9. Berman NS (1986) "Velocity fluctuations in non-homogeneous drag reduction" Chem Eng Commun 42: 37
10. Berner C, Scrivener O (1980) "Drag reduction and structure of turbulence in dilute polymer solutions" Prog. Astronaut. Aeronaut., Vol 72 Viscous flow drag reduction (AIAA), 290
11. Bewersdorff HW (1982) "Effect of centrally injected polymer thread on drag in pipe flow" Rheol Acta 21: 587
12. Bewersdorff HW (1984) "Heterogene Widerstandverminderung bei turbulenten Rohrströmungen" Rheol Acta 23: 522
13. Bewersdorff HW (1986) „Widerstandsverminderung in turbulenten Rohrströmungen" Proc Hamburger Makromol Symp
14. Bewersdorff HW, Frings B, Lindner P, Oberthuer RC (1986) "The conformation of drag reducing micelles from small-angle-neutron-scattering experiments" Rheol Acta 25: 642
15. Bewersdorff HW, Berman NS (1987) "Effect of roughness on drag reduction for commercially smooth pipes" J. Non-Newton Fluid Mech 24: 365
16. Bewersdorff HW, Singh RP (1987) "Turbulent drag reduction and relaminarisation by xanthan gum" IUTAM-Symposium, Bangalore, India
17. Bird RB, Armstrong RC, Hassager O (1977) "Dynamics of polymeric liquids" Vol 1: Fluid Mechanics, J. Wiley, New York
18. Bird RB, Hassager O, Armstrong RC, Curtiss CF (1977) "Dynamics of polymeric liquids" Vol 2: Kinetic theory, J. Wiley, New York
19. Black TJ (1969) "Viscous drag reduction examined in the light of a new model of wall turbulence" in "Viscous drag reduction" ed Wells CS, Plenum Press, New York, p. 383
20. Blatch NS (1906) "Water filtration at Washington" Trans Am Soc Civ Eng 57: 400
21. Bock J, Kowalik RM, Valint PL (1986) "Drag reduction agent" U.S. Pat. 4585810A
22. Boog U, Tianbao H, Meyerhoff G (1987) "The sedimentation behaviour of semi-dilute solutions of poly(methyl methacrylate)" Eur Polym J 23: 781
23. Boggs FW, Tompson J (1966) "Flow properties of dilute solution of polymers" Part I: Mechanism of drag reduction U.S. Rubber Co. Research Center Report on Contract No. Nonr. 3123(00)
24. Bouldin M, Kulicke W-M, Kehler H (1988) "A prediction of the non-Newtonian viscosity and shear stability of polymer solutions" Colloid Polym Sci 266: 793
25. Brandrup J, Immergut EH (1975) "Polymerhandbook" Chap. IV, J. Wiley, New York
26. Brautlecht CA, Sethi JR (1933) "Flow of paper pulps in pipe lines", Ind Eng Chem 25: 283
27. Brennen C (1970) "Some cavitation experiments with dilute polymer solutions" J Fluid Mech 44: 51
28. Brostow W, Macip MA, Sochanski JS (1982) "Macromolecular conformations in solutions" J Statist Phys 29: 849
29. Brostow W (1983) "Drag reduction and mechanical degradation in polymer solutions in flow" Polymer 24: 631
30. Brostow W, Ertepinar H (1984) "Chain semiflexibility and drag reduction in polymer solution flow" Polym Prepr (Am Chem Soc Div Polym Chem) 25: 69
31. Brunn PO (1987) "Some modern developments in the flow of dilute polymer and surfactant solutions" Physico Chem Hydrodyn 8: 449
32. Burger ED, Chorn LG, Perkins TK (1980) "Studies of drag reduction conducted over a broad range of pipeline conditions when flowing Prudhoe Bay crude oil" J Rheol 24: 603
33. Carville DB, Feig ER, Rose GD, Pulliam MH (1986) "Additives for non-aqueous liquids" U.S. Pat. 454573488A
34. Chang HFD, Meng JS (1987) "Prediction of drag reduction from rheological properties of dilute polymer solutions" Physicochem Hydrodyn 9: 33
35. Chao KK, Child CA, Grens EA, Williams MC (1984) "Antimisting action of polymeric additives in jet fuels" AIChE J 30: 111
36. Corino ER, Brodkey RS (1969) "A visual investigation of the wall region in turbulent flow" J Fluid Mech 37: 1
37. Cottrell FR, Merrill EW, Smith KA (1970) "Intrinsic viscosity and axial extensions ratio of random-coiling macromolecules in a hydrodynamic shear field" Polymer 8: 287

38. Daoud M, Cotton JP, Farnoux B, Jannink G, Sarma G, Benoit H, Dupressix R, Picot C, DeGennes PG (1975) "Solutions of flexible polymers. Neutron experiments and interpretation" Macromolecules 6: 804
39. DeGennes PG (1979) "Scaling concepts in polymer physics" Cornell Univ. Press, Ithaca, New York
40. Dembek G, Bewersdorff HW (1981) „Kurzfristige Kapazitätserhöhung von Abwässerkanälen durch Zusatz von wasserlöslichen Polymeren" gwf-Wasser/Abwasser 122, H 9: 392
41. Deshmukh SR, Chaturvedi PN, Singh RP (1985) "The turbulent drag reduction by graft copolymers of guar gum and polyacrylamide" J Appl Polym Sci 30: 4013
42. Deshmukh SR, Singh RP (1986) "Drag reduction characteristics of graft copolymers of xanthangum and polyacrylamide" J Appl Polym Sci 32: 6163
43. Deshmukh SR, Singh RP (1987) "Drag reduction effectiveness, shear stability and biodegradadation resistance of guargum-based graft copolymers" J Appl Polym Sci 33: 1963
44. Dodge DW, Metzner AB (1959) "Turbulent flow of non-Newtonian system" AIChE J 5: 189
45. Dschagarowa E, Mennig G (1976) „Verminderung des Reibungswiderstandes von Flüssigkeiten im turbulenten Bereich mittels hochpolymerer Zusätze" Fortschritte VDI-Zeitschriften 7: 41
46. Dschagarowa E, Mennig G (1977) "Influence of molecular weight and molecular conformation of polymers on turbulent drag reduction" Rheol Acta 16: 309
47. Dschagarowa E, Bochossian T (1978) "Drag reduction in polymer mixtures" Rheol Acta 17: 426
48. Dschagarowa E, Bochossian T (1980) „Der Einfluß des Kapillarmaterials auf die Verringerung des Reibungswiderstandes" Rheol Acta 19: 258
49. Dunlop EH, Cox LR (1977) "Influence of molecular aggregates on drag reduction" Phys Fluids 20, Part II: 203
50. Dunn P, Oldfield D (1970) "Tri-n-butyl-tin-fluoride. A novel coordination polymer in solution" J Macromol Sci A4: 1157
51. Durst F, Rastogi AK (1977) "Calculations of turbulent boundary layer flows with drag reducing polymer additives" Phys Fluids 20: 1975
52. Durst F, Haas R, Interthal W, Keck T (1982) „Polymerwirkungen in Strömungen — Mechanismen und praktische Anwendungen" Chem Ing Tech 54: 213
53. Durst F, Haas R, Interthal W (1982) "Laminar and turbulent flows of dilute polymer solutions: a physical model" Rheol Acta 21: 572
54. Durst F, Schmitt K (1985) „Einsatz von reibungsvermindernden Additiven in Fernwärmenetzen. Hydromechanische Grundlagenuntersuchungen" Int Report, University Erlangen-Nürnberg
55. Durst F, Interthal W (1985) "The nature of flow through porous media" in "The influence of polymer additives on velocity and temperature fields" ed Gampert B, Springer-Verlag, Berlin, New York
56. Ellis AT, Ting RY, Napolink RH (1970) "Some effects of storage and shear history on the friction reducing properties of dilute polymer solutions" Prog Astronaut Aeronaut 70: Viscous flow flow drag reduction (AIAA) 532
57. El'perin IT, Smol'skii BM, Leventhal LI (1967) "Decreasing the hydrodynamic resistence of pipelines" Int Chem Eng 7: 276
58. Evans AP (1974) "A new drag-reducing polymer with improved shear stability for non-aqueous systems" J Appl Polym Sci 18: 1919
59. Fabula AG, Lumley JC, Taylor WD (1966) "Some interpretations of the Toms-effect" in "Modern development in the mechanics of continua" ed Eskinazi S, Academic Press, New York, p. 145
60. Fabula AG (1971) "Fire fighting benefits of polymeric friction reduction" Trans ASME, J Basic Eng 453
61. Falco RE, Wiggert DC (1980) "Effects of dilute polymer solutions in vortexring/wall interactions — a mechanism of drag reduction" Prog Astronaut Aeronaut 72: Viscous flow drag reduction (AIAA) 275
62. Ferry JD (1980) "Viscoelastic properties of polymers" 2nd ed J. Wiley, New York
63. Forame PC, Hansen RJ, Little RC (1972) "Observations of early turbulence in the pipe flow of drag reducing polymer solutions" AIChE J 18: 213

64. Forrest F, Grierson GAH (1931) "Friction losses in cast iron pipe carrying paper stock" Paper Trade J 298
65. Fortuin JMH, Klijn PJ (1982) "Drag reduction and random surface renewal in turbulent pipe flow" Chem Eng Sci 37: 611
66. Fortuna G, Hanratty TJ (1972) "The influence of drag reducing polymers on the viscous sublayer" J Fluid Mech 53: 575
67. Frings B (1985) "Annular injection of concentrated polymer solutions into the wall region of a turbulent pipe flow" in "The influence of polymer additives and velocity and temperature fields" ed Gampert B, Springer-Verlag, Berlin, New York
68. Frings B (1988) "Heterogeneous drag reduction in turbulent pipe flows using various injection techniques" Rheol Acta 27: 92
69. Gadd GE (1965) "Turbulence damping and drag reduction produced by certain additives in water" Nature 206: 463
70. Gadd GE (1966) "Differences in normal stress in aqueous solutions of turbulent drag reducing additives" Nature 212: 1348
71. Gadd GE (1966) "Reduction of turbulent friction in liquids by dissolved additives" Nature 212: 874
72. Gampert B, Wagner P (1982) „Experimentelle Untersuchungen turbulenter Rohrströmungen hochreiner wässriger Polyacrylamidlösungen" Rheol Acta 21: 578
73. Gampert B, Wagner P (1982) "Turbulent flow with polymer additives" Arch Mech 34: 493
74. Gampert B, Wagner P (1982) "Reduced turbulence production by increased elongational viscosity" "Recent contribution to fluid mechanics" ed Haase W, Springer, Berlin-New York
75. Gampert B, Wagner P (1985) "The influence of molecular weight and molecular weight distribution on drag reduction and mechanical degradation in turbulent flow of highly dilute polymer solutions" in "The Influence of Polymer Additives on Velocity and Temperature Fields" ed Gampert B, Springer-Verlag, Berlin, New York
76. Giesekus H, Bewersdorff HW, Dembeck G, Kwade M, Martischius FD, Scharf R (1981) „Rheologie" Fortschritte Verfahrenstechnik 19: 3
77. Giesekus H, Bewersdorff HW, Frings B, Hibberd M, Kleinecke K, Kwade K, Möller D, Schröder R (1985) „Rheologie" Fortschritte Verfahrenstechnik 23: 3
78. Gold PI, Amar PK, Swaidan BE (1973) "Friction reduction degradation in dilute poly(ethylene oxide) solutions" J Appl Polym Sci 17: 333
79. Golda J (1986) "Hydraulik transport of coal in pipes with drag reducing additives" Chem Eng Comm 43: 53
80. Gordon RJ (1970) "On the explanations and correlations of turbulent drag reduction in dilute macromolecular solutions" J Appl Polym Sci 14: 2097
81. Gordon RJ, Balakrishnan C (1972) "A new viscoelastic effect with importance in drag reduction and polymer characterization" J Appl Polym Sci 16: 1629
82. Goren Y, Norbuary IF (1967) "Turbulent flow of dilute aqueous polymer solutions" Trans ASME, J Basic Eng 89: 814
83. Gramain P, Borreill J (1978) "Influence of molecular weight and molecular structure of polystyrenes on turbulent drag reduction" Rheol Acta 17: 303
84. Graessley W (1980) "Polymer chain dimensions and the dependence of viscoelastic properties on concentration, molecular weight, and solvent power" Polymer 21: 258
85. Greene HL, Mostardi RF, Wokes RF (1980) "Effects of drag reducing polymers on initiation of arteriosclerosis" Polym Eng Sci 20: 499
86. Gundert F, Wolf BA (1986) "Viscosity of dilute polymer solutions: molecular weight dependence of the Huggins coefficient" Makromol Chem 187: 2969
87. Gyr A (1968) "Analogy between vortex stretching by drag-reducing additives and vortex stretching by fine suspensions" Nature 219: 928
88. Gyr A, Mueller A (1974) "The effect of wall adsorption on the Toms effect" Chem Eng Sci 29: 1057
89. Gyr A, Mueller A (1975) "Alteration of structures of sublayer flow in dilute polymer solutions" Nature 253: 185
90. Hanratty TJ, Eckelmann LD, Fortuna G (1972) "Drag reduction and the wave length of flow-oriented wall eddies" Nature Phys Sci 236: 94

91. Hashemzadeh A, Kulicke W-M (1986) „Degradation von Polymermolekülen beim Durchströmen poröser Medien" Chem Ing Tech 58: 325
92. Haas R, Kulicke W-M (1984) "Flow behavior of dilute polyacrylamide solutions through porous media" II. Indirect determination of extremely high molecular weights and some aspects of viscosity decrease over long time intervals Ind Eng Chem Fundam 23: 316
93. Hershey HC, Zakin JL (1967) "Existence of two types of drag reduction in pipe flow of dilute polymer solutions" Ind Eng Chem Fundam 6: 381
94. Hlavacek B, Rollin LA, Schreiber HP (1976) "Drag reduction effectiveness of macromolecules" Polymer 17: 81
95. Hoyt JW, Fabula AG (1964) "The effect of additives on fluid friction" Proc 5th Symp on Naval/ Hydrodynamics, Bergen, Norway 112: 947
96. Hoyt JW (1971) "Drag-reduction effectiveness of polymer solutions in the turbulent-flow rheometer: a catalog" J Polym Sci, Polym Lett 9: 851
97. Hoyt JW (1971) "Blood transfusion fluids having reduced turbulent friction properties" U.S. 3.590.124
98. Hoyt JW (1972) "The effect of additives on fluid friction" Trans ASME, J Basic Eng 94: 258
99. Hoyt JW (1974) "Recent progress in polymer drag reduction" Colloques Internationaux du C.N.R.S. No. 233 — Polymers et lubrification 193
100. Hoyt JW (1985) "Drag reduction in polysaccharide solutions" Trends in Biotechnol 3: 17
101. Hoyt JW (1986) "Drag reduction" Encycl Polym Sci Eng 5: 129, ed Kroschwitz JI, J. Wiley, New York
102. Hunston DL, Griffith JR, Little RC (1973) "Drag reducing properties of polyphosphates" Nature Phys Sci 246: 140
103. Interthal W (1984) „Wasserlösliche Hochpolymere als Strömungsbeschleuniger" Angew Makromol Chem 123/124: 387
104. Interthal W, Wilski H (1985) "Drag reduction experiments with very large pipes" Colloid Polym Sci 263: 217
105. Ishibashi O, Ihara M (1984) "Drag coefficient of a setting particle in dilute polymer solutions" Nippon Kogyo Kaishi 100: 29
106. James DF, Saringer JH (1980) "Extensional flow of dilute polymer solutions" Fluid Mech 97: 655
107. James DF, McLean BD, Saringer JH (1987) "Presheared extensional flow of dilute polymer solutions" J Rheol 31: 453
108. Johnson B, Barchi RH (1968) "Effect of drag reducing additives on boundary-layer turbulence" J Hydronautics 2: 108
109. Keller A, Mackley MR (1977) "The possible relevance of persistent extensional flow on the interpretation of drag reduction phenomena" 2nd Intern Conf on drag reduction, paper F 1, Cambridge, England
110. Kemmer FN (1979) "The NALCO Water Handbook" McGraw-Hill, New York
111. Kenis PR, Hoyt JW (1971) "Friction reduction by algal and bacterial polymers" J Fluid Mech 50: 133
112. Kenis PR (1971) "Turbulent flow friction reduction effectiveness and hydrodynamic degradation of polysaccharides and synthetic polymers" J Appl Polym Sci 15: 607
113. Kim OK, Little RC, Ting RY (1973) "Polymer structural effects in turbulent drag reduction" AIChE Symp Series 69: 39
114. Kim OK, Long T, Brown F (1986) "An unusual coil deformation of polyacrylic acid induced by high shear flow" Polymer Commun 27: 71
115. Kilian FP (1970) „Über die Verminderung des Reibungswertes von Grenzschichtströmungen viskoelastischer Flüssigkeiten" Mitteilung der Versuchsanstalt für Wasserbau und Schiffbau Berlin 52
116. Kinnier JW (1970) "A correlation between friction reduction and molecular size for the flow of dilute aqueous polyethyleneoxide solutions in pipes" Thesis, Y. S. Naval Postgraduate School
117. Klein J, Conrad KD (1978) "Molecular weight determination of polyacrylamide and polyacrylamide-co-sodium acylate" Makromol Chem 179: 1635
118. Kohn MC (1973) "Energy storage in drag-reducing polymer solutions" J Polym Sci Polym Phys Ed 11: 2339

119. Kowalik RM, Duvdevani I, Peiffer DG, Lundberg RD, Kitano K, Schulz DN (1987) "Enhanced drag reduction via interpolymer associations" J Non-Newton Fluid Mech 24: 1
120. Kulicke W-M, Kniewske R (1980) "Purification of aqueous and non-aqueous polymer solutions for light scattering measurements and a comparison of molecular weights determined with a low-angle and a wide-angle-photometer" Makromol Chem, Rapid Commun 1: 719
121. Kulicke W-M, Kniewske R, Klein J (1982) "Preparation, characterization, solution properties and rheological behaviour of polyacrylamide" Prog Polym Sci, 8: 373
122. Kulicke W-M, Elsabee M, Eisenbach CD, Peuscher M (1983) "Effect of the molecular weight and molecular weight distribution on the rheological properties of aqueous polyethylenoxide solution" Polym Bull 9: 190
123. Kulicke W-M, Böse N (1984) „Bestimmung der Molmassenverteilung sowie der Stabilitätsgrenze von Polyacrylamiden unter Benutzung einer kombinierten Ausschlußchromatographie/Kleinwinkel-Laser-Lichtstreuphotometer-Anlage" Colloid Polym Sci 262: 197
124. Kulicke W-M, Haas R (1984) "Flow behaviour of dilute polyacrylamide solutions through porous media" I. Influence of chain length, concentration and thermodynamic quality of the solvent Ind Eng Chem Fundam 23: 308
125. Kulicke W-M, Klein J (1985) "Molecular parameters as a basis of structure property relations in the flow of polymer solutions" in "The influence of polymer additives and velocity and temperature fields" ed Gampert B, Springer-Verlag, Berlin-New York
126. Kulicke W-M, Hörl HH (1985) "Preparation and characterization of a series of poly(acrylamide-co-acrylates), with a copolymer composition between 0–96.3 mol% acrylate units with the same degree distribution of polymerization" Colloid Polym Sci 263: 530
127. Kulicke W-M (ed) (1986) „Fließverhalten von Stoffen und Stoffgemischen" Hüthig & Wepf, Heidelberg
128. Kulicke W-M (1986) "Unusual instability effects observed in ionic and non-ionic water soluble polymers" Makromol Chem, Macromol Symp 2: 137
129. Kulicke W-M, Kniewske R, Müller R-J, Prescher M, Kehler H (1986) „Scherung und Degradation von Polymerlösungen" Angew. Makromol Chem 142: 29
130. Kulicke W-M (1987) "Prediction of viscosity level in the non-Newtonian range and shear stability of polymer fluids" Proceed 2nd Plym Symp, 42, Napa Valley, USA
131. Kulicke W-M, Kehler H, Bouldin M "A consideration of the state of solution in the molecular modeling of the zero-shear viscosity for polymeric liquids" Colloid Polym Sci (submitted)
132. Kulicke W-M, Böse N, Bouldin M (1988) "The role of polymers in enhanced oil recovery" Contribution to the book "Water-soluble polymers for petroleum recovery" eds Schulz G, Stahl A, Plenum Press (in press)
133. Kuo Y, Tanner RI (1974) "Use of open-channel flows to measure the second normal stress differences" Rheol Acta 13: 931
134. Lacey PMC (1974) "Drag reduction by long-chain polymers" Chem Eng Sci 29: 1495
135. Lagerstedt T, Alfredsson PH "A note on the drag reducing properties of fish mucus" Technical Report TRITA-MEK-83-11, ISSN 0348-U67X Royal Institute of Technology, Stockholm
136. Latto B, Shen CH (1970) "Effect of dilute polymer solutions injection on external boundary layer phenomenon" Can J Chem Eng 48: 34
137. Layec Y, Layec-Raphalen MN (1983) "Instability of dilute poly(ethylenoxide) solutions" J Phys Lett 44: 121
138. Leca M, Leca A (1985) "Reduction of pressure loss in liquid flow by means of additives" V. Dependence of efficiency on the method of synthesis degree of polyacrylamides. Rev Chim 36: 936
139. Lechner MD, Mattern R (1984) „Das Verhalten von Polyvinylalkohol in wässriger Lösung" Angew. Makromol Chem 123/124: 43
140. Little RC (1967) "Drag reduction by dilute polymer solutions in turbulent flow" Naval Research Laboratory Report 6542
141. Little RC (1969) "Displacement of aqueous drag-reducing polymer solutions" Ind Eng Chem Fundam 8: 520
142. Little RC (1969) "Flow properties of polyox solutions" Ind Eng Chem Fundam 8: 557
Little RC, Hansen RJ, Hunston DL, Kim OK, Patterson RL, Ting RY (1975) "The drag reduc-

tion phenomenon. Observed characteristics, improved agents and proposed mechanismus" Ind Eng Chem Fundam 14: 283

144. Lockett FJ (1964) "Fluid dynamics approach to the Toms effect" Nature 222: 937
145. Lumley JL (1964) "Turbulence in non-Newtonian fluids" Phys Fluids 7: 335
146. Lumley JL (1967) "The Toms phenomenon: Anomalous effects in turbulent flows of dilute solutions of high molecular weight linear polymers" Appl Mech Rev 20: 1139
147. Lumley JL (1969) "Drag reduction by additives" Ann Rev Fluid Mech 1: 367
148. Lumley JL (1973) "Drag reduction in turbulent flow by polymer additives" J Polym Sci Macromol Rev 7: 263
149. Lumley JL (1977) "Drag reduction in two phases and polymer flows" Phys Fluids 20: 10
150. Malhotra JP, Deshmukh SR, Singh RP (1987) "Turbulent drag reduction by polymer-fiber mixtures" J Appl Polym Sci 33: 2467
151. Martischius F-D, Münstedt H (1985) „Widerstandsverminderung in turbulenten Rohrströmungen: Stabilitätsverhalten wässriger Polyacrylamidlösungen bei mechanischer und thermischer
152. Beanspruchung" Rheol Acta 24: 92
Maus JR, Wilhelm LR (1970) "Effect of polymer injections on frictional drag in turbulent pipe flow" J Hydronautics 4: 35
153. McComb WD, Chan KTJ (1981) "Drag reduction in fibre suspension" Nature 292: 520
154. McComb WD, Rabie LH (1982) "Local drag reduction due to injection of polymer solutions into turbulent flow in a pipe" AIChE J 28: 547
155. McCormick CL, Hester RD, Morgan SE, Safieddine AM (1987) "Model water-soluble acrylamide copolymers for the study of polymeric drag reduction" Polym Mater Sci Eng 57: 840
156. Metzner AB, Metzner AP (1970) "Stress levels in rapid extensional flows of polymeric fluids" Rheol Acta 9: 174
157. Metzner AB, Kale DD (1976) "Turbulent drag reduction in dilute fiber suspensions: Mechanistic considerations" AIChE J 22: 669
158. Metzner AB (1977) "Polymer solution and fiber suspension rheology and their relationship to turbulent drag reduction" Phys Fluids 20: 145
159. Mysels (1949) "Flow of thickened fluids" US Pat 2, 492, 173 (Dec. 27)
160. Nagarajan R (1971) "Studies on non-Newtonian liquids" Thesis, Madras, India
161. Ogawa K, Kuroda C (1986) "Experimental study on the effect of elasticity on drag reduction and turbulent fluctuations in the laminar-turbulent transition region in pipe flow of dilute polymer solutions" Can J Chem Eng 64: 497
162. Ohlendorf D, Interthal W, Hoffmann H (1986) "Surfactant systems for drag reduction: physicochemical properties and rheological behaviour" Rheol Acta 25: 468
163. Oldroyd JG (1948) "Suggested method of detecting wall effects on turbulent flow throught tubes" Proceed 1st Intern Congr on Rheology II, 130, North Holland, Amsterdam
164. Oliver DR, Bakhtiyarov SI (1983) "Drag reduction in exceptionally dilute polymer solutions" J Non-Newton Fluid Mech 12: 113
165. Onogi S, Masuda T, Yanaga N, Kimura Y (1967) "Dependence of viscosity of concentrated polymer solutions upon molecular weight and concentration" J Polym Sci Part A-2, 5: 899
166. Parker CA, Hedley AH (1972) "Drag reduction and molecular structure" Nature Phys Sci 236: 27
167. Patterson GK, Zakin JL (1968) "Prediction of drag reduction with a viscoelastic modell" AIChE J 14: 434
168. Patterson GK, Zakin JL, Rodriguez JM (1969) "Drag reduction. Polymer solutions, soap solutions and solid particle suspensions in pipe flow" Ind Eng Chem 61: 22
169. Patterson RW, Abernathy FH (1970) "Turbulent flow drag reduction and degradation with dilute polymer solutions" Fluid Mech 43: 689
170. Peterlin A (1970) "Molecular model of drag reduction by polymer solutes" Nature 227: 598
171. Peyser P, Little RC (1971) "The drag reduction of dilute polymer solutions as a function of solvent power, viscosity and temperature" J Appl Polym Sci 15: 2623
172. Peiffer DG, Lundberg RD, Kowalik RM, Turner SR (1984) "Drag reduction agents for aqueous salt solutions" U.S. Pat 4460758
173. Pfenninger W (1967) "A hypothesis of the reduction of the turbulent friction drag in fluid flows by means of additives" Northrop Corp Norair Division Report BLC-179
174. Povkh IL, Chernyuk VV (1986) "Experimental study of the effect of polyacrylamide additions on the flow resistance in diffusers" Inzh-Fiz Zh 51: 357

175. Ram A, Finkelstein E, Elata C (1967) "Reduction of friction in oil pipelines by polymer additives" Ind Eng Chem Proc Des Devel 6: 309
176. Ramu KLV, Tullis JP (1976) "Drag reduction and velocity distribution in developing pipe flow" J Hydronautics 10: 55
177. Reddy GV, Singh RP (1985) "Drag reduction effectiveness and shear stability of polymer-polymer and polymer-fiber mixtures in recirculatory turbulent flow of water" Rheol Acta 24: 296
178. Reddy GV (1986) "Drag reduction by polymers in recirculatory flow" Chem Eng World 21: 73
179. Rochefort S, Middleman S (1985) "Effect of molecular configuration of xanthan gum drag reduction" in "Polymer-Flow Interaction" ed Rabin Y, AIP Conf Proc 137: 117
180. Rosen MW, Cornford NE (1971) "Fluid friction of fish slimes" Nature 234: 49
181. Rossetti SJ, Pfeffer R (1972) "Drag reduction in dilute flowing gas-solid suspensions" AIChE J 18: 31
182. Ruckenstein E (1973) "A note on the mechanism of drag reduction" J Appl Polym Sci 17: 3239
183. Rudd MJ (1972) "Velocity measurements made with a laser dopplermeter on the turbulent pipe flow of dilute polymer solutions" Fluid Mechan 51: 673
184. Rudd MJ (1971) "Laser dopplermeter and polymer drag reduction" Chem Eng Prog Symp Ser 111: 67, 21
185. Savins JG (1964) "Drag reduction characteristics of solution of macromolecules in turbulent pipe flow" Soc Petrol Eng J 4: 203
186. Schmerwitz H, Reher EO (1986) "Effect of elastic properties on turbulent flow of petroleum and petroleum products in pipelines" Wiss Z Tech Hochsch "Carl Schorlemmer" Leuna-Merseburg 28: 235
187. Schulz DN, Peiffer DG, Kowalik RM, Kaladas JL (1985) "Drag reduction agent" U.S. Pat 4560710A
188. Sellin RHJ (1974) "The suppression of turbulent by drag reducing polymer additives" Colloq Intern CRNS No 233 Brest
189. Sellin RHJ, Hoyt JW, Scrivener O (1982) "The effect of drag reducing additives on fluid flows and their industrial applications. Part 1: Basic aspects" J Hydraulic Res 20: 29
190. Sellin RHJ, Hoyt JW, Pollert J, Scrivener O (1982) "The effect of drag reducing additives on fluid flows and their industrial applications" Part 2: Present applications and future proposals J Hydraulic Res 20: 235
191. Shaver RG, Merril EW (1959) "Turbulent flow of pseudoplastic polymer solutions in straight cylindrical tubes" AIChE J 5: 181
192. Shenoy AV (1984) "A review on drag reduction with special reference to micellar systems" Colloid Polym Sci 262: 319
193. Shin H (1965) "Reduction of drag in turbulence by dilute polymer solutions" Thesis, MIT, USA
194. Simha R, Utracki LA (1973) "The viscosity of concentrated polymer solutions: corresponding states principles" Rheol Acta 12: 455
195. Smith JM, Bartels PV (1985) "Influence of drag reducing agents on mixing in turbulent pipe flow" Paper presented on the workshop "Rheology" of the professional commitee held in Bonn, Feb. 25
196. Stenberg LG, Lagerstedt T, Lindgren ER (1977) "Polymer additives mixing and turbulent drag reduction" Phys Fluids 20: 276
197. Thorne PF (1974) "Drag reduction in fire fighting" Proc Int Conf in drag reduction, Cambridge, UK, H1
198. Tiederman WG, Luchik TS, Bogard DG (1985) "Wall-layer structure and drag reduction" J Fluid Mech 156: 419
199. Ting RY, Kim OK (1972) "Drag reduction properties of ultrahigh molecular weight polyacrylamide and related polymers" Am Chem Soc Div Org Coating Plastics Chem 32: 270
200. Toms BA (1948) "Some observations on the flow of linear polymer solutions through straight tubes at large Reynolds-numbers" Proc 1st Intern Congr on Rheology North Holland Amsterdam II: 135
201. Townsend AA (1956) "The structure of turbulent shear flow" University Press, New York
202. Tulin MP (1966) "Hydrodynamic aspects of macromolecular solutions" Proceed 6th Symp on Naval Hydrodynamics, ONR CR-136, p 3, Washington
203. Tullis JP, Ramu KLV (1974) "Drag reduction in developing pipe flow with polymer injection" Intern Conf on Drag Reduction, paper G 3, Cambridge

204. Usui H, Kodama M, Sano Y (1987) "Drag reduction with heterogeneous polymer injection into a turbulent pipe flow" Technol Rep Yamaguchi Univ 4: 51
205. Virk PS (1967) "The Toms-phenomenon: turbulent pipe flow of dilute polymer solutions" Fluid Mech 30: 305
206. Virk PS (1971) "Drag reduction in rough pipes" Fluid Mech 45: 225
207. Virk PS (1975) "Drag reduction fundamentals" AIChE J 21: 625
208. Vleggaar J (1973) "Studies on drag reduction" Thesis, University Amsterdam
209. Vleggaar J, Tels M (1973) "Drag reduction by polymer threads" Chem Eng Sci 28: 965
210. Vleggaar J, Tels M (1973) "Heat transfer in a heterogeneous drag reduction system" Int J Heat and Mass Transfer 16: 1629
211. Wagner P (1984) „Der Einfluß molekularer Parameter auf das Verhalten turbulent strömender Polymerlösungen" Thesis, University Essen
212. Walsh M (1967) "Theory of drag reduction in dilute high-polymer flows" Int Shipbuilding Prog 14: 134
213. Walters RR, Wells CS (1971) "An experimental study of turbulent diffusion of drag reducing polymer additives" J Hydronautics 5: 65
214. Warner HR (1972) "Kinetic theory of dilute suspensions of finitely extendible dumbbells" Ind Eng Chem Fundam 11: 379
215. Wells CS, Spangler JG (1967) "Injection of a drag reducing fluid into turbulent pipe flow of a Newtonian fluid" Phys Fluids 10: 1890
216. Wolff C (1982) "Non-Newtonian behaviour of associations of macromolecules in dilute solutions" Adv Coll Interf Sci 17: 263
217. Wunderlich AM, James DF (1987) "Extensional flow resistance of dilute polyacrylamide and surfactant solutions" Rheol Acta 26: 522
218. Zakin JL, Chang JC (1972) "Nonionic surfactants as drag reducing additives" Nature Phys Sci 239: 26

Application of NMR to Crosslinked Polymer Systems

Mladen Andreis[1] and Jack L. Koenig
Department of Macromolecular Science, Case Western Reserve University
Cleveland, OH 44106, USA

Approximately 80% of synthetic polymers for commercial use are crosslinked. Yet, there are very few techniques which reveal the nature of the crosslink structure of these polymers. This article reviews the utility of NMR for the characterization of the type, number and distribution of crosslinks in polymeric materials. It discusses the theory and experimental ramifications of this unusual application of NMR. This is the first article directed specifically towards developing an understanding of this aspect of NMR. It summarizes all aspects of solution and solid state NMR including chemical shift and relaxation measurements for the study of crosslinking. Due to the recent developments in solid state NMR and deuterium NMR, our understanding of the crosslinked structure of polymers has been substantially increased through NMR. It is important that an appreciation of the developing potential of high resolution solid state NMR be recognized so it can be used in the future.

[1] Permanent address: Ruder Boskovic Institute, Zagreb, Yugoslavia

Advances in Polymer Science 89

1 Introduction

The application of nuclear magnetic resonance (NMR) spectroscopy to polymer systems has contributed to significant advances in understanding of their structure and dynamical properties at the molecular level. From the analytical point of view, NMR spectroscopy is particularly suitable for a determination of the polymer structure by direct observation of the protons and carbons in different structural moieties. However, until the mid-1970s the application of this technique was limited to polymer solutions and to some elastomers in the solid state with a relatively high degree of the molecular mobility which allows the observation of the motionally narrowed absorption signals.

The importance of crosslinked polymers, since the discovery of cured phenolic formaldehyde resins and vulcanized rubber, has significantly grown. Simultaneously, the understanding of the mechanism of network formation, the chemical structure of crosslinked systems and the motional properties at the molecular level, which are responsible for the macroscopic physical and mechanical properties, did not accompany the rapid growth of their commercial production. The insolubility of polymer networks made impossible the structural analysis by NMR techniques, although some studies had been made on the swollen crosslinked polymers.

In the early stages of the development of NMR techniques (1950s–mid 1970s), the studies of polymers could be classified into two major domains: broad line NMR of the solid state and the high resolution NMR of polymer liquids and solutions. In this period, crosslinked polymers were investigated by the broad line and pulsed NMR techniques, respectively. These studies in the solid state yield information primarily on macromolecular dynamics, and indirectly on the network structure.

With the development of Fourier transform (FT) techniques in NMR spectroscopy (early 1970s), the first major advance in the NMR technology was made. A significant increase in the sensitivity, as compared to the conventional continuous wave method, resulted in the NMR spectroscopy of rare nuclei, particularly ^{13}C NMR, which is essential for polymer studies. The ^{13}C NMR analysis of swollen lightly crosslinked polymers was made possible. The relaxation measurements, based on the different pulse sequences, provided additional information on the network dynamics.

A direct method for the determination of the chemical structures of crosslinked polymers became available in the mid-1970s with the second major advance in NMR spectroscopy — high resolution of solids. The combination of different techniques, such as dipolar decoupling (DD), magic angle spinning (MAS) and cross-polarization (CP) makes possible the observation of the chemical structures of polymers in the solid state. With this improvement, the growth of solid state techniques, as well as their application to crosslinked polymers, became very rapid. A number of published NMR papers indicate the complexity of the crosslinking mechanism and the network structures and dynamics in the solid state.

The importance of NMR spectroscopy in determining the polymer molecular structure and dynamics, as well as the rapid development of spectroscopic techniques, resulted in a number of review articles, which have appeared since the late 1950s [1–8]. The particular types of polymers, as well as the particular NMR rechniques are separately reviewed, such as the characterization of crosslinked polymers by high resolution solid state NMR [9].

The scope of the paper is to review the application of different NMR techniques, particularly the high resolution solid state methods, to crosslinked polymers and the potential of each particular technique in the investigation of network structure and dynamics at the molecular level.

2 Principles of NMR Spectroscopy

Nuclear magnetic resonance (NMR) is based on a phenomenon that nuclei which possess both magnetic and angular moments (i.e. have odd mass number or odd atomic number) interact with an applied magnetic field $\mathbf{B}_0$ yielding 2I + 1 (where I is the nuclear spin quantum number) energy levels with separation ΔE:

$$\Delta E = h\omega = \gamma h B_0 \tag{1}$$

where h is the Planck constant divided by 2π, ω is the Larmor frequency of nuclear precession and γ is the gyromagnetic ratio.

The interaction of a single spin with the magnetic field (in the range 10^6 to 10^8 Hz) is described by a Zeeman Hamiltonian:

$$\mathbf{H}_Z = -\gamma h B_0 \mathbf{I}_z \tag{2}$$

where $\mathbf{I}_z$ is the z-component of the spin angular momentum operator **I** (in the direction of the applied field). Spectroscopy detection of these energy levels is possible when transitions between them are induced by an alternating magnetic field $B_1(\omega t)$ of the frequency ω (perpendicular to the static field B_0) which satisfy the resonance condition $\omega = \gamma B_0$.

The properties of multi-spin systems are determined by different types of interactions. They can be described by a Hamiltonian **H**:

$$\mathbf{H} = \mathbf{H}_Z + \mathbf{H}_D + \mathbf{H}_Q + \mathbf{H}_J + \mathbf{H}_\delta \tag{3}$$

where Hamiltonian terms are described as follows:

$\mathbf{H}_Z$ = Zeeman interaction with the applied field;
$\mathbf{H}_D$ = direct dipole-dipole interaction with other nuclei;
$\mathbf{H}_Q$ = quadrupolar interaction (for nuclei with $I > 1/2$);
$\mathbf{H}_\delta$ = chemical shift interaction;
$\mathbf{H}_J$ = indirect (electron-coupled) spin-spin couplings to other nuclei.

Contributions of the four last terms in Eq. (3), depend on the physical state. In the solid state, the strong dipolar and quadrupolar terms are usually dominant, and the weak interactions such as chemical shift and spin-spin coupling are obscured.

In contrast, the dipolar and quadrupolar interactions in liquids are averaged to zero, giving rise to high resolution spectra in which chemical shifts and J-couplings can be observed. Furthermore, rapid motions in solutions average the above mentioned tensors to scalar quantities. According to these observations, NMR studies can be classified into two major domains: broad line (low resolution) studies of the solid state and sharp line (high resolution) studies in the liquid state.

However, it is found that a combination of techniques, such as proton dipolar decoupling (removes the dipolar interactions), magic angle spinning (reduces the chemical shift tensor to the isotropic chemical shift value), and cross-polarization (increases the sensitivity of rare spins, like ^{13}C) applied to a solid state material, results in sharp lines for ^{13}C nuclei in the solid state [10]. Thus, the observation of narrow lines or high resolution NMR in the solid state is possible.

3 Solid State NMR

As previously observed, the dominant terms in the Hamiltonian which describe a spin system in the solid state are the dipolar and quadrupolar terms. In the case of nuclei with I = 1/2 (such as ^{1}H, ^{13}C, ^{19}F and ^{29}Si) the quadrupolar interaction is zero. The dipolar Hamiltonian $\mathbf{H}_D$ (for a homonuclear spin system) has the general form:

$$\mathbf{H}_D = \gamma^2 h^2 \sum_{i<j} \left[\frac{\mathbf{I}_i \mathbf{I}_j}{\mathbf{r}_{ij}^3} - \frac{3(\mathbf{I}_i \mathbf{r}_{ij})(\mathbf{I}_j \mathbf{r}_{ij})}{\mathbf{r}_{ij}^5} \right] \tag{4}$$

where γ is the gyromagnetic ratio of the nuclei at the resonance; $\mathbf{I}_i$ and $\mathbf{I}_j$ are the nuclear spin angular momentum operators, and $\mathbf{r}_{ij}$ is the vector joining the ith and jth nuclei.

Two types of low resolution solid state NMR techniques can be distinguished: a) broad line NMR in which the absorption signal is obtained by sweeping the magnetic induction B in the vicinity of the resonance value B_0 (Eq. (1)), and b) pulsed techniques which are based on the possibility of rotating the magnetization under the influence of particular radiofrequency pulses, or pulse sequences.

3.1 Broad Line NMR

The proton NMR absorption signal of a solid homogeneous polymer usually consists of a single broad peak. This is a consequence of a large number of interactions between the various nuclear magnetic moments which give rise to a local field, $\mathbf{B}_{loc}$, contributing to the single resonance field, $\mathbf{B}_0$. Since almost every proton exists in a slightly different magnetic environment, the resonance envelope consists of a superposition of numerous individual resonances generating a single broad absorption line. Theoretical calculations of $\mathbf{B}_{loc}$ and their contributions to the shape of the resonance lines are possible only for the simplest molecules. Thus, the more suitable quantity, the moment of the absorption line, ΔM_n, based on a quantum mechanical determination of the mean square local field in the rigid lattice, was introduced [11]. The n-th moment is defined in terms of the magnetic field as:

$$\Delta M_n = \frac{\int_{-\infty}^{+\infty} (B - B_0)^n \, I(B) \, d(B - B_0)}{\int_{-\infty}^{+\infty} I(B) \, d(B - B_0)} \tag{5}$$

where I(B) is the shape function of the resonance absorption signal. In broad line spectroscopy, the NMR spectrum is recorded as a first derivative, dI/dB. Particularly, the second moment of the NMR line, ΔM_2 (n = 2), analogous to the moment of inertia, is the frequently used moment. For the spin system consisting of N spins of spin I, it is related to the internuclear distance r_{ij}:

$$\Delta M_2 = \frac{3}{4} \gamma^4 h^2 I(I + 1) \frac{1}{N} \sum_{i,j} \frac{(1 - \cos^2 \theta_{ij})^2}{r_{ij}^6} \tag{6}$$

where θ_{ij} is the angle between r_{ij} and the direction of applied magnetic field B_0. Another important parameter in broad line NMR is line width (expressed in terms of frequency), $\Delta\nu$, where $\nu = \omega/2\pi$. A line width in terms of magnetic field (frequently denoted as δH) is defined as the difference between two field values in the derivative spectrum, B_a and B_b, which have maximum and minimum spectral intensity, respectively.

Variations in broad line NMR are primarily sensitive to motion. Consequently, changes in $\Delta\nu$ and ΔM_2 are indications of motional processes in the system. If, as a first approximation, the relaxation process can be expressed by a single correlation time, τ_c, the following relation between NMR parameters and τ_c can be established:

$$\tau_c = \frac{\tan [\pi(\Delta\nu)^2/2(\Delta\nu_{RL})^2]}{\alpha \, \Delta\nu} \tag{7}$$

where $\Delta\nu_{RL}$ is the line width of the rigid lattice (absence of motion on NMR time scale), and α is a line shape parameter [12].

Under the assumption of only one temperature dependent relaxation process, the correlation time (as a quantitative measure of molecular motion) is related to the activation energy, E_a, by the Arrhenius equation:

$$\tau_c = \tau_0 \exp (E_a/RT) \tag{8}$$

where τ_0 is the time factor and R is the gas constant. It should be noticed that the activation energies determined from NMR experiments are generally lower than the E_a for the same processes calculated from dielectric or mechanical relaxations. The assumption of a single correlation time for motional processes in polymers does not generally give a satisfactory explanation of the complex macromolecular dynamics. However, when a single τ_c and the corresponding E_a (as relative quantitative measures of motional processes) are observed one obtains significant information on polymer dynamics.

Early investigations of crosslinked polymeric materials observed line width changes with cure. An increase in the line width is observed in vulcanized elastomers at room temperature, compared with the unvulcanized samples [13]. The effect of vulcanization on line widths of butadiene-styrene copolymers in a temperature range near the glass transition (T_g), was interpreted in terms of the formation of cooperative domains, which lead to larger linewidths ($\Delta\nu$) [14]. Both, an increase in $\Delta\nu$ and shift of the transition range to higher temperatures due to the restriction of molecular mobility induced by crosslinks, are generally observed in cured elastomers [15,16], and also in ^{19}F NMR of irradiated polytetrafluoroethylene [17].

Linewidth vs temperature curves for vulcanized natural rubber exhibit two different transition regions. The low temperature $\Delta\nu$ change (shifted to higher temperatures with the increasing curing time) is attributed to methyl group reorientation, and the second higher more abrupt temperature change to segmental motion (Fig. 1). Increasing temperatures lead to the onset of methyl group and segmental motions, respectively, and corresponding activation energies calculated from Eqs. (7) and (8) indicate that the methyl group motion is more affected by vulcanization than the segmental motion (Table 1) [18].

Greater degree of curing (longer curing times and higher amounts of sulfur) shifts the transition region to higher temperatures. The effect of vulcanization on line shape is rather complex, and at least one additional important parameter (crystallinity) should be considered [19]. The fact that uncured natural rubber exhibits line narrowing at somewhat higher temperatures than vulcanized rubbers with low sulfur content, suggests that crosslinks reduce the crystallinity, and restrictions of molecular motion induced by crosslinking are less effective than those imposed by the crystalline regions. However, constraints due to the crosslinking became dominant at higher sulfur content. Analogous effects on crystalline and amorphous regions are found in polypropylene irradiated with increasing doses [20]. The temperature dependence of $\Delta\nu$ and ΔM_2 for cis-1,4-polybutadiene indicate that the differences between mobili-

Table 1. Apparent activation energies for relaxation processes in crosslinked polymers calculated from broad line NMR

Polymer	Measure of crosslinking	E_a/kJ mole^{-1}	
		Low temperature process	High temperature process
Cured natural rubber[a]	30[d]	11.3[g]	53.6
	60	17.2	54.4
	90	18.8	55.6
Polydimethylsiloxane[b]	0[e]		36.4
	3		33.5
	6		15.9
	25		9.6
	60		9.2
	600		8.8
Polydimethylsilazane[c]	Linear polymer	94.6[h]	8.4
	1:1	11.7	94.6
	3:1	9.6	107.9

[a] Taken from Ref. [18]; [b] taken from Ref. [38]; [c] taken from Ref. [40]; [d] curing time, min; [e] ratio of multifunctional monomers methyltrichlorosilane and dimethyldichlorosilane; [f] irradiation dose in Mrep (1 rep, i.e. roentgen-equivalent-physical is the dose of ionizing radiation which produces an energy absorption of 93 erg/cm^3 (alternatively 84 erg/cm^3) in the tissue; this unit is sometimes also expressed in terms of energy absorption per gram of tissue. Hence, it cannot be consistently converted into the SI system of units (Gy). Given doses in Mrep are an approximate measure for the applied irradiation. [g] Methyl group rotation; [h] chain bending or stretching

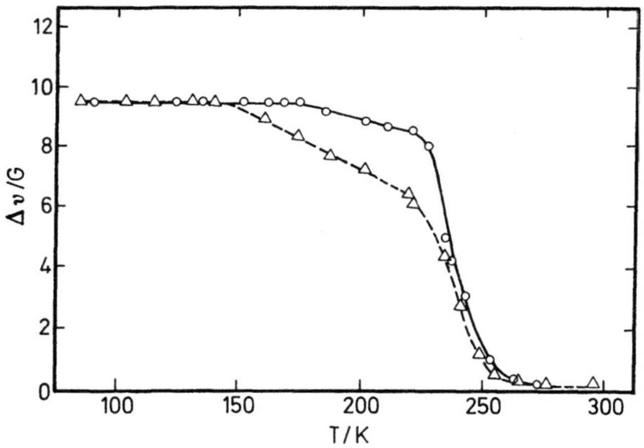

Fig. 1. Temperature dependence of the proton resonance line width (Δν) in vulcanized natural rubber cured for 30 (△) and 90 (○) minutes (adapted from Ref. [18])

ties of protons in amorphous and crystalline regions in the course of vulcanization are reduced, and the system becomes more homogeneous. However, the aging process in both a filled rubber and a vulcanizate, gives rise to increasing regions with higher mobility [21].

Broad line NMR spectra of heterogeneous systems, such as crystalline or crosslinked polymers exhibit a complex line shape consisting of two maxima which can be separated in components (broad and narrow) [22] for certain ratios of the widths and compositions [23]. Areas under these curves are proportional to the number of protons in each phase. However, quantitative analysis requires: 1) very accurate determination of line shapes (Lorentzian, Gaussian, or the more general, Weibull function) because of the significant differences in calculated areas (about 60% between Lorentzian and Gaussian curves with the same width and intensities), and 2) precise decomposition of overlapping spectra, particularly in the case of superimposed narrow Lorentzian lines. Lorentzian lines are not generally suitable for precise calculation of moments because broad "wings", which contain a lot of information, must be cut at a certain level.

Line shapes for polymers vary from the Gaussian type at low temperatures to the Lorentzian type at higher temperatures. However, crosslinked polymers, such as styrene-divinyl benzene copolymers, poly(dimethyl siloxane) and unsaturated polyesters, exhibit both components in the intermediate temperature range corresponding to the glass transition temperature. The temperature, where the narrow non-Gaussian line begins to appear, is shifted to higher temperatures with the increasing degree of crosslinking. The appearance of two-component spectra in the transition region is caused by differences in segmental motions in polymer networks [24]. The more mobile motions result in the narrow component being observed. The narrow component appears at higher temperatures with increasing filler content in filled cured epoxy resins [25].

Irradiated polyethylene exhibits different effects on the two observed broad and narrow components [26,27,28,29]. The first effect of irradiation on motional behavior (when the dose reaches a critical value) is reduction of crystallinity which increases the amorphous volume and, consequently the intensity of the narrow line component. However, as the dose is increased the more dominant effect is the restriction of segmental and chain motions in amorphous regions due to the crosslinking (Fig. 2). An investigation of γ-irradiation on crystalline and amorphous regions in poly(vinyl alcohol) in the air and in vacuum was also made [30].

The superposition of two lines, correponding to protons of different mobility, and the temperature dependence of ΔM_2 observed in unsaturated polyesters cured with different amounts of polystyrene, indicate the presence of two thermal transitions in these polymer networks [31]. Network formation in the temperature induced curing of epoxy-resin-modified poly(ester amide) changes the NMR spectrum to a broad line (protons in cured region), on which is superimposed a narrow line (protons in uncured region). The fraction of uncured resin is determined from the calculated areas under corresponding line components. Furthermore, from the changes in two superimposed NMR lines, it is concluded that the additional curing (increasing intensity and $\Delta\nu$ of broad line) in the first stage of the aging process of epoxy-resin-modified poly(ester amide) at higher temperatures, is followed by a scission of the molecules (decreasing $\Delta\nu$ of narrow line) in the second stage of aging [32]. The kinetics of unsaturated polyester hardening was also followed using the areas of broad and narrow lines in connection with spin-spin relaxation times [33].

Irradiation of butadiene rubbers gives rise to a broad component, with the intensity increasing with the irradiation time. It is found that rubbers crosslinked with irradiation have more hindred motion than rubbers crosslinked with sulfur; this is explained by the higher potential barrier of rotation about C—C bonds, as compared to C—S bonds [34].

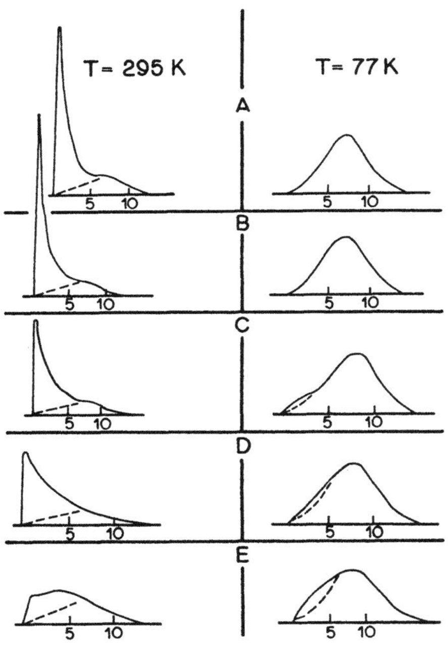

Fig. 2. Effect of reactor irradiation on ^{1}H NMR derivative line shapes of polyethylene at 295 and 77 K (only the positive half of line shape is shown). Irradiation doses (10^{18} nvt): 0 (A), 0.3 (B), 0.6 (C), 1.2 (D) and 8.0 (E) (adapted from Ref. [28])

Two component spectra of plasma polymerized styrene (PPS) observed above 50 °C were interpreted in terms of a mesh-like crosslinked network structure. Broad and narrow components are attributed to the protons in crosslinked regions and oligomeric structures (which act as plasticizers), respectively [35].

Investigations of the curing of unsaturated poly(ester maleates/fumarates) with styrene and triethyleneglycol dimethylacrylate oligomer (TGM-3), show that the process of cold hardening after the gel-formation reduces the molecular mobility during the first 10 days (increase of ΔM_2); in the time interval from 10 to 60 days, the ΔM_2 is practically unchanged. Theoretical values of intramolecular components of the ΔM_2 for TGM-3, and styrene crosslinked resins (calculated from ΔM_2 of individual groups) are 11.4 and $12.2 \cdot 10^{-8}$ T^2, respectively. Considerably higher experimental values for TGM-3 cured resins ($16.9 \cdot 10^{-8}$ T^2) are explained in terms of additional network structural units and the possible presence of the homopolymer of the oligomer TGM-3 [36].

An increase of $\Delta\nu$ to a constant value was observed in the course of vulcanization of polysulphide rubbers in the presence of sodium dichromate. Rates of vulcanization were determined from the time dependence of $\Delta\nu$, as the reciprocal time at which $\Delta\nu$ reaches the maximum constant value. It is found that the rate of vulcanization linearly increases with the oxidizer concentration [37].

Irradiated polydimethylsiloxane gums (PDMS) exhibit a lowering of the transition temperatures and increasing of $\Delta\nu$ at room temperature, with the increasing dose and crosslinking, for the samples with relatively small crosslink densities [38]. Lower transition temperatures are accompanied with decreasing activation energies, as calculated from the linear dependence of τ_c vs. $1/T$ (Table 1). The decrease in E_a is explained in terms of loss in long-range order, i.e. loss in the degree of cooperativeness necessary for rotational freedom. The observed increase of $\Delta\nu$ at high temperatures is interpreted in terms of hindered translational motion. Furthermore, it is shown that different types of crosslinking in nitrogen containing dimethylpolysilazanes have different effects on molecular motions [39]. Dimethylpolysilazanes crosslinked through silicon show more diffuse transitions, shifted to higher temperatures with the introduction of crosslinks. Moreover, plots of τ_c vs. $1/T$ for both, linear and crosslinked samples, exhibit a curvature which cannot be fitted with a single straight line and hence, a single E_a. From the limiting maximum and minimum slope of each curve, a range of plausible activation energies has been determined (Table 1). The high temperature barrier, which sharply increases with the crosslinking, is assigned to the onset of translational motions of the polymer chains; and the low temperature barrier, sharply decreasing with the crosslinking, is assigned to local chain bending or stretching modes restricted by the defects introduced by crosslinks [40].

The temperature dependence of the second moment for crosslinked epoxy polymers [based on I) diglycidyl ether of resorcinol and II) 2,6-diaminopyridine, modified by a monofunctional epoxy monomer, phenylglycidyl ether (PGE)], indicates three thermal transitions: a) a high temperature α-process, which is analogous with the glass transition, but in polymer networks is probably related to rotational freedom in the segments between crosslinks; b) the β-transition ascribed to the motion of the hydroxyether fragments, and c) the low temperature γ-process, probably related to the motion of components of the polymer chain (rotation of benzene rings). The temperatures of all three transitions increase with the increasing crosslink density, but ΔM_2 at low temperatures is smaller for a highly crosslinked network as compared with the less crosslinked samples [41].

On the contrary, higher ΔM_2 at any measured temperature is observed for highly crosslinked epoxies based on I) prepolymers of bisphenol A-type with varying chain lengths and II) 4,4'-diaminodiphenylmethane (DDM). At the same time, the line width, $\Delta\nu$, below the transition temperature is found to be the same for uncured and cured epoxies, indicating that the motions of small groups within the repeat unit are not affected by crosslink density. The activation energy for this motional process is 7.1 kJ/mol; the apparent E_a for the α-process (including energy required for the formation of free volume) is about 146 kJ/mole [42].

It seems that the effect of crosslinking on low temperature motions is specific for particular crosslinked polymers. Three general types of $\Delta\nu$ and ΔM_2 dependence in the low temperature region, for uncrosslinked and crosslinked polymers, respectively, can be observed. A few additional examples will illustrate this behavior. a) ΔM_2 values at 77 K are higher for poly(vinyl trichloroacetate) (PVTAc) crosslinked with poly(methyl methacrylate) (PMMA) than for uncrosslinked PMMA; simultaneously, the transition range associated with the development of the α-methyl group rotation can occur at either lower, equal or higher temperatures, compared to the uncrosslinked PMMA. Although this behavior cannot be quantitatively interpreted in terms of molecular structure (crosslink density), it seems that enhancement of the methyl-group rotation arises in the mixed PVTAc-PMMA phase, and the retardation of the rotation in the PMMA phase. Generally, higher ΔM_2 values may arise from the unusual conformations adopted by the PMMA chains and modified intra- and inter-molecular contributions to the second moment [43]. b) $\Delta\nu$ and ΔM_2 for irradiation crosslinked PDMS at 90 K are independent of dose, indicating that the glass structure is not significantly changed by crosslinking up to a ratio of about one crosslink for every five monomer units [38]; the limiting value at low temperatures (rigid lattice line width, $\Delta\nu_{RL}$) is also found to be the same for all sulfur vulcanized natural rubber samples, cured for 30, 60 and 90 min, respectively [18]. c) The pretransition $\Delta\nu$ of dimethylpolysilazanes decreases with increased crosslinking through silicon due to the lower concentration of methyl groups within the sample which results in less intermolecular broadening and a narrower line [40].

Proton and fluorine are the most frequently studied nuclei in the solid state NMR of polymers. However, useful information can be obtained from deuterium resonance, (^2H; $I = 1$). In the presence of nuclei with spin $I > 1/2$, which possess a quadrupolar moment, the most dominant interaction occurs between the quadrupolar moment and an electric field gradient tensor **V**, generated by the C—D bonding electrons. Consequently, the only dominant interaction, besides $\mathbf{H}_Z$ (Eq. (1)), will be described by the quadrupolar Hamiltonian, which for a single spin has the form:

$$\mathbf{H}_Q = \mathbf{IQI} \tag{9}$$

where **Q** is the quadrupolar tensor. For an unoriented amorphous sample, the NMR spectrum will consist of a so-called "doublet" structure, where the two lines of major intensity are separated by the quadrupole splitting $\Delta\nu_Q = (3/4)\chi$ (χ is the nuclear quadrupole coupling constant). The main advantage of using deuterium is the specific sensitivity to molecular motions, when the deuterium is introduced in selected locations in the molecule. Motions which are fast compared to the quadrupole interaction modulate V and narrow the spectrum. The type of motion present in polymers produces for ^2H NMR studies predictable changes in the line shapes.

Deuterium broad line NMR is used in the study of curing of phenolic resins. In order to investigate both crosslinking bridges and the aromatic rings, two separate deuterated samples are prepared: [D_2]formaldehyde and [3,5-D_2]phenol. Deuterium NMR spectra for both components recorded for the resins are shown in Fig. 3. Figure 3A shows that slightly different uncured prepolymers are formed from the reaction of phenol with formaldehyde. The degree of crosslinking is obtained from methylene/methylol ratio by high resolution ^{13}C NMR and it is 0.30 and 0.35 for spectrum A and B, respectively. The [D_2]formaldehyde spectra exhibit only a narrow line, indicating a high degree of mobility in the prepolymer; the two component spectrum of [3,5-D_2]phenol with a broad quadrupole pattern suggests that despite the presence of a narrow line, a large portion (~50%) of the prepolymer is already crosslinked.

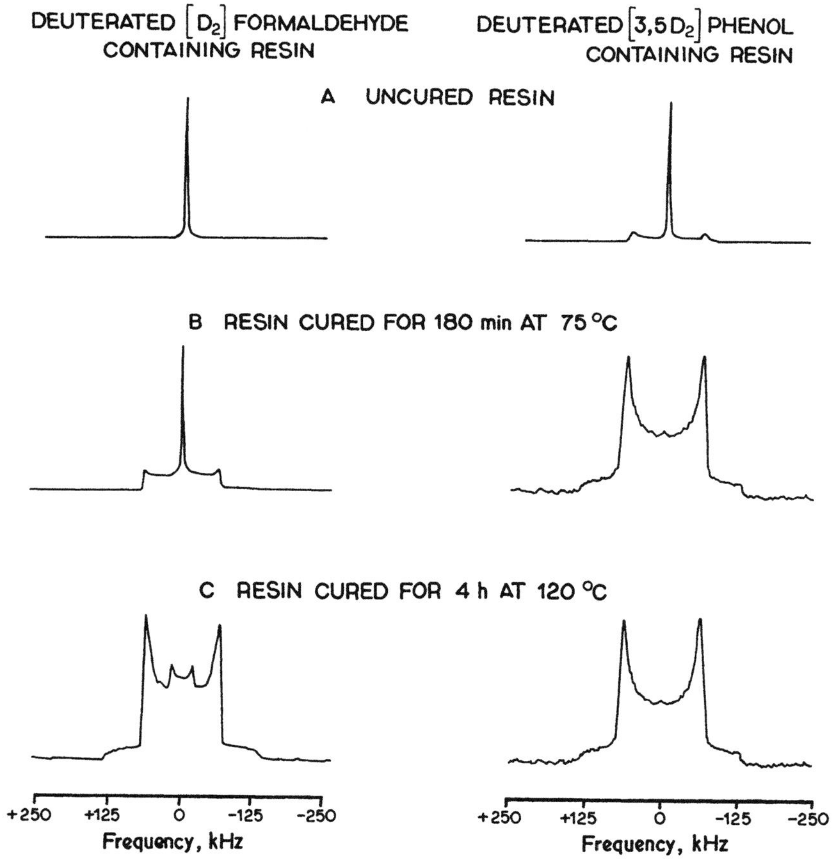

Fig. 3. ^2H NMR spectra of uncured (A) and (B, C) cured phenol/formaldehyde resins: (I) deuterated [D_2]formaldehyde containing resin and (II) deuterated [3,5-D_2]phenol containing resin (adapted from Ref. [44])

However, these differences in prepolymers have no significant influence on the dynamic structure of the cured resins. The [3,5-D_2]phenol part of the resin cured at 75 °C does not exhibit the narrow isotropic components, indicating that phenyl rings are completely rigid (probably due to the crosslinking at the 2, 4 and 6 positions of the ring). The [D_2]formaldehyde part of the spectrum shows a decrease in the intensity of the narrow line and the appearance of the quadrupole pattern (restriction in motions); however, the narrow isotropic line, in cured resins, seems to be the result of small deuterium containing molecules in the curing reaction (for example CD_2O), rather than an indication of rapid motions of the polymer molecules. The narrow component disappears in the resin cured at 120 °C and a new quadrupolar pattern ($\Delta\nu_Q = 40$ kHz) appears. This pattern is ascribed to the rotation of CD_3 groups, formed as a degradation product; methyl groups are not found in the phenolic counterpart of the spectrum [44].

Deuterium NMR is very sensitive to orientational behavior and order; there are a number of papers dealing with constrained polymeric networks. For example, ^{2}H NMR (in both, solid state and solution) is used in the study of the orientational order generated in uniaxially strained rubbers as a function of the crosslink density. Two sets of rubbers (model end-linked silicone rubbers and randomly crosslinked diene networks) were investigated directly (on perdeuterated silicone labelled chains) and indirectly, via C_6D_6 as an NMR probe for diene rubbers [45].

3.2 Pulsed NMR

Molecular motions in low molecular weight molecules are rather complex, involving different types of motion such as rotational diffusion (isotropic or anisotropic torsional oscillations or reorientations), translational diffusion and random Brownian motion. The basic NMR theory concerning relaxation phenomena (spin-spin and spin-lattice relaxation times) and molecular dynamics, was derived assuming Brownian motion by Bloembergen, Purcell and Pound (BPP theory) [46]. This theory was later modified by Solomon [46] and Kubo and Tomita [48]; an additional theory for spin-lattice relaxation times in the rotating frame was also developed [49].

The frequency ranges covered by these three relaxation times; spin-lattice (T_1); spin-spin (T_2), and spin-lattice in the rotating frames ($T_{1\varrho}$), are determined by the following equations:

$$\begin{aligned} 1/T_1 &\propto J_1(\omega_0) + J_2(2\omega_0) \\ 1/T_{1\varrho} &\propto J_0(2\omega_1) + J_1(\omega_0) + J_2(2\omega_0) \\ 1/T_2 &\propto J_0(0) + J_2(\omega_0) + J_2(2\omega_0) \end{aligned} \tag{10}$$

where J_is are the spectral density functions of the general form:

$$J_i(\omega) = \frac{1}{2}\int_{-\infty}^{+\infty} G_i(\tau)\, e^{i\omega t}\, d\tau \tag{11}$$

i.e. $J_i(\omega)$ is the frequency spectrum corresponding to the autocorrelation function $G(\tau)$:

$$G_i(\tau) = \overline{\langle F_i(\tau)\, F_i^*(t + \tau)\rangle} \tag{12}$$

or its Fourier transform at the resonance frequency ω_0. Spectral density functions J_i ($i = 0, 1, 2$), correspond to cases in which there are no net flips, one flip and two flips between spins I_1 and I_2, respectively. As can be seen from Eq. (10), the J_is which determine the T_1, are functions of the rf field ω_0 and, hence, T_1 values depend on the resonance frequency $\omega_0 = \gamma B_0$. In contradistinction, dominant contributions to the T_2 relaxation time in solids arise from the frequency independent term $J_0(0)$ and, consequently, the temperature of abrupt change in T_2, which follows a phase transition, is not frequency dependent. The $T_{1\varrho}$ relaxation time is predominantly governed by frequency of the spin-locking field ω_1.

The sensitivity of the three relaxation times to the molecular dynamics and structure will be discussed in a subsequent section. The general temperature dependence of T_1, $T_{1\varrho}$ and T_2 for a typical linear amorphous polymer with one side group attached to a backbone is shown in Fig. 4.

3.2.1 Spin-Lattice Relaxation Times

The relationship between spin-lattice relaxation time (T_1) and $J_i(\omega_0)$, can be derived through the time dependent perturbation theory by calculating transition probabilities

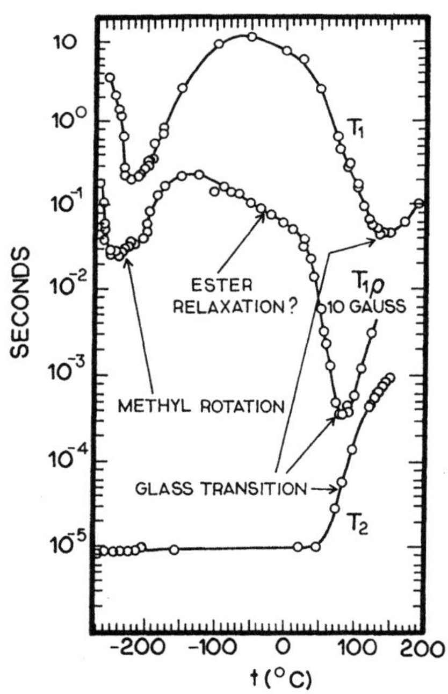

Fig. 4. Pulsed NMR data for poly(vinyl acetate), as a typical example for polymer relaxation. The glass transition corresponds to the abrupt change in T_2 and the high temperature minima in T_1 and $T_{1\varrho}$ curves. Lower temperature minima in T_1 and $T_{1\varrho}$ curves are attributed to a side group relaxation (reprinted from Ref. [1]) with permission)

between energy levels in a model spin system consisting of a pair of spins. In the case of spins with $I = 1/2$, the following equation is established:

$$\frac{1}{T_1} = \frac{3\gamma^4 h^2}{10} \sum_j r_j^{-6}[J(\omega_0) + 4J(\omega_0)] \tag{13}$$

where γ is the nuclear gyromagnetic ratio and r_j is the distance between neighboring nuclei and ω_0 is the angular resonance frequency ($2\pi\nu_0$).

The simplest model, for the autocorrelation function $G(\tau)$ commonly used in other relaxation processes, is single exponential:

$$G(\tau) = \exp(-|\tau|/\tau_c) \tag{14}$$

where τ_c denotes corrrelation time. Using this model with the single correlation time, Eq. (13) can be written as:

$$\frac{1}{T_1} = \frac{3\gamma^4 h^2}{10} \sum_j r_j^{-6} \left[\frac{\tau_c}{1 + \omega_0^2\tau_c^2} + \frac{4\tau_c}{1 + 4\omega_0^2\tau_c^2}\right] \tag{15}$$

In limiting cases $\omega_0\tau_c \ll 1$ and $\omega_0\tau_c \gg 1$, $1/T_1$ is directly proportional to τ_c and $1/\tau_c$, respectively. A plot of log T_1 vs. $1/T$ is a V-shaped curve with two straight line components and the predicted minimum occurs when $\omega_0\tau_c = 0.6158$.

The measurement of the longitudinal (spin-lattice) relaxation time T_1 is based on the application of appropriate pulse sequence and monitoring the free induction decay following the sequence. Initially, the pulse sequences were introduced in pulsed NMR for the measurement of basic relaxation times, while the broad line and high resolution techniques were performed by the continuous wave method, i.e. sweeping the magnetic induction $\mathbf{B}_0$ in the vicinity of the resonance frequency ω_0. However, with the development and application of Fourier-transform (FT) techniques in magnetic resonance, the pulse methods became the basic of all modern FT NMR high resolution techniques in both solution and solid state. A large number of pulse sequences were developed, resulting in a variety of new dynamical and structural information. All pulsed experiments are based on the different responses of the magnetization, presented for simplicity by a classical vector rather than a quantum-mechanical quantity, to the perturbations caused by the particular pulses. As an example, the behavior of magnetization under the simple pulse sequences, which explains the nature of all pulsed measurements, is shown for the relaxation in the static magnetic field ($\mathbf{B}_0$) and the rotating frame ($\mathbf{B}_1$), respectively (Figs. 5 and 8).

The most commonly used pulse sequence for the T_1 measurement is the inversion-recovery sequence ($180° - \tau - 90° - t_d$), shown in Fig. 5 [50]. The 180° pulse inverts the magnetization $\mathbf{M}_0$ along the $-z$ axis ($\mathbf{M}_z = -\mathbf{M}_0$): the longitudinal relaxation process now occurs, causing the return of the time dependent magnetization $-\mathbf{M}_0$ through zero to its equilibrium value $\mathbf{M}_0$. This process is defined by the spin-lattice relaxation time T_1. The application of a second pulse after the time τ, i.e. the 90° degree pulse along the x axis, causes the rotation of the magnetization $\mathbf{M}(\tau)$ to the y axis, which is sampled by the detector. A free induction decay (FID) results,

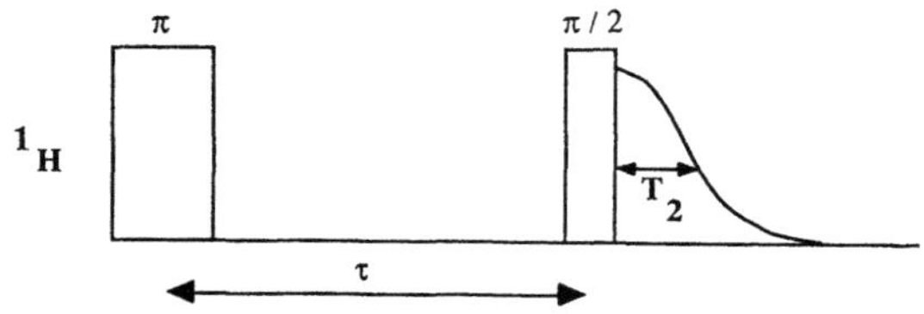

Fig. 5. ^1H T_1 pulse sequence

with the intensity proportional to the value of **M** at the time τ. After the delay time which is equal to at least 5 times T_1 the $\mathbf{M}(\tau)$ will practically return to the equilibrium $\mathbf{M}_0$ (99.3% of its value). By changing the values of τ and repeating the pulse sequence, the T_1 values are calculated from the exponential plot of $\mathbf{M}(\tau)$ vs. τ.

T_1 times are most sensitive to high-frequency molecular motions in the range of the NMR frequency ($\nu \sim 10^7$ Hz); minima in T_1 vs. temperature curves correspond to the particular relaxation processes in the system: high temperature relaxation processes in amorphous polymers are generally connected to the glass transition, and minima, in a lower temperature range to the relaxations involving reorientations of the side groups (Fig. 4).

The spin-lattice relaxation process is usually exponential. Theoretically, the effect of spin-diffusion, characterized by the coefficient D (order of 10^{-12} cm^2 s^{-1}), has an influence on T_1 relaxation times when $T_1 > L^2/D$, where L is the diffusion path length. NMR studies of model systems for rubber networks, based on a styrene-butadiene-styrene block copolymer (SBS), in which styrene blocks act as a crosslink for polybutadiene rubber segments of known and uniform length, indicate that spin diffusion operating between PS and PB phases causes a lowering of T_g for the PS component in SBS (as compared to the pure PS) and hindering of the motion of the PB component (as compared to the pure PB) [51].

Studies of epoxy-rubber cured with a diamine over a wide temperature range, indicate a complex behavior of the spin-lattice relaxation time. In the temperature range 163 to 243 K, a single T_1 is observed, due to the rapid spin-diffusion between coupled protons of rubber and the polyepoxide phase. At higher temperatures (253 to 313 K), two T_1 times of the order of 10^{-2} s can be resolved: the longer (T_{1L}) and the shorter (T_{2S}), are attributed to the protons in epoxide and rubber phases, respectively, (fraction of protons having T_{1S} relaxation times corresponds to the calculated number of protons in rubber). It can be assumed that in this temperature range there is no detectable interaction between phases. At 343 K, again a single T_1 is observed; at 433 K there is a second phase separation with T_{2L} arising from rubber molecules, and T_{2S} from an epoxy phase [52].

However, a single T_1 is generally observed in heterogeneous systems; partly because it is difficult to resolve T_1 components unless they differ by at least a factor of two. Deuterium T_1 relaxation times in crosslinked resins, also exhibit only a single-exponential relaxation process, indicating that different ^2H in the same molecule have essentially the same spin-lattice relaxation times [44].

Relaxation studies on natural rubber (uncrosslinked and crosslinked) [19] show that the experimental T_1 curve is narrower near the T_1 minimum than predicted by the BPP theory. Although the depth of the T_1 minimum in some polymers, for example, the one due to the methyl group reorientation in uncured diglycidyl ether of bisphenol-A (DGEBA), is found to agree with the value predicted by the BPP theory. The T_1 minima in cured resins are broader and shallower [53]. In general, experimental data for macromolecules exhibit deviations from the BPP theory (disagreement) is greater in the case of crosslinked polymers). It is observed that high- and low-temperature slopes of the log T_1 vs $1/T$ plot in uncrosslinked polybutadiene are not equal, due to entanglements which cause motional constraints; there is also doubt if the calculated E_a, using Eq. 8, in these cases really represents the activation energy [54].

In order to obtain better agreement with experimental results, the concept of a distribution of correlation times was introduced in nuclear magnetic relaxation. Different distribution functions, $G(\tau_c)$, such as Gaussian, and functions proposed by Yager, Kirkwood and Fuoss, Cole and Cole, and Davidson and Cole (asymmetric distribution) are introduced into the Eq. (13), giving a general expression for T_1 [55]:

$$\frac{1}{T_1} = C\left[\int_0^\infty \frac{\tau_c G(\tau_c)\, d\tau_c}{1 + \omega_0^2\tau_c^2} + 4\int_0^\infty \frac{\tau_c G(\tau_c)\, d\tau_c}{1 + 4\omega_0^2\tau_c^2}\right] \tag{16}$$

where the constant C contains γ, h and r, and the normalized distribution function must satisfy:

$$\int_0^\infty G(\tau_c)\, d\tau_c = 1 \tag{17}$$

Schaefer suggested a form of the $G(\tau_c)$ function more suitable for macromolecules by using the asymmetrical distribution functions χ^2 and $\log \chi^2$, with the higher density in the range of longer correlation times [56]. Other modifications of the BPP theory include substitution of exponential correlation functions with the nonexponential form in terms of two correlation times (defect diffusion model [57] and a diamond lattice model (VJGM; three- and four-bond jumps on a diamond lattice [58]) and as a product of three correlation functions connected with the particular types of motions in polymer melts [59]. A modification of this model which is based on locyl defects associated with the reptation model [60] is used in the line shape analysis of swollen crosslinked gels. Howarth's 3τ libration-rotation model [61], which is based on Woessner's anisotropic 2τ rotation model [62] and Howarth's 2τ libration model [63], describes molecular motions with multiple correlation times. This model assumes two librational modes of motion at different rates and amplitudes, superimposed on a very slow isotropic motion. A discrete model based on a three-bond crankshaft motion, characterized by a single exponential τ_c and Woessner's internal anisotropic motion were also proposed [64], as well as versions extended in the continuum limit [not restricted to a lattice] [65,66].

Rowland and Labun developed a model which is suitable for crosslinked polymers using a Gaussian distribution of τ_c. This model is based on spin diffusion to locations of rapid spin-lattice relaxation and assumes that protons in the crosslink region have different relaxation times (T_{1x}) from those in the free chain region (T_{1c}) [54]:

$$\frac{1}{T_1} = \frac{f_x}{T_{1x}} + \frac{1 - f_x}{T_{1c}} \tag{18}$$

where f_x is the mole fraction of repeat units in the crosslinks. Using the appropriate equation derived from Eq. (15) (assuming Gaussian distribution of τ_c), experimentally observed T_1 data on lightly crosslinked polybutadiene are found to be in good agreement with the proposed model (Fig. 6). Analysis of the data for networks with sufficient distances between crosslinks ($f_x < 0.4$) suggests that: a) crosslinks may be viewed as independent entities, b) the distribution τ_c associated with the crosslink region is very broad, and c) the relaxation behavior of the chain segments in a network is well described by the relaxation of the uncrosslinked polymer.

However, there is still controversy in the literature concerning the effect of crosslinking on T_1 relaxation times. Vulcanization of natural rubber causes an increase in the temperature of the T_1 minimum with increasing crosslink density, indicating more hindered motions. Simultaneously, the apparent activation energies of the relaxation process are smaller, as calculated from the temperature dependence of T_1 and Eq. (8) (Table 2) [19]. This behavior is ascribed to the complex frequency distribution of the lattice motions. It is concluded that lower values of E_a coupled with higher values of T_{1min} are indicative of a broader distribution of correlation times [67]. Decreasing of E_a with the increase in crosslinking, is interpreted in terms of increasing distances between the polymer chains with the introduction of cross-

However, an increase in both $T_{1\,min}$ and in the temperature at which it occurs, observed in crosslinked polyurethanes with increasing crosslink density, does not
links which provide more space for segmental motion [68,69].
affect the corresponding activation energies [70]. Moreover, the sensitivity of temperature and time values at $T_{1\,min}$ to the reduction of polyurethane crosslink density, imposed by the monofunctional monomers (free chains), and to the hydrogen bonds present in the system, are found to be dependent on the particular polyurethane systems [70,71].

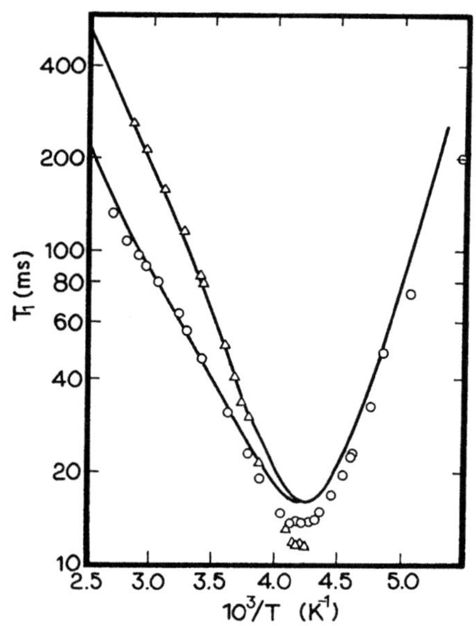

Fig. 6. Proton T_1 relaxation data for crosslinked polybutadiene samples with average number of repeat units per network chain 44 (△) and 14 (○), compared with the computed results from the modified BPP equation (assuming Gaussian distribution and the model based on spin diffusion to locations of rapid spin-lattice relaxation) (reprinted from Ref. [54] with permission)

Table 2. Apparent activation energies for relaxation processes in crosslinked polymers calculated from T_1 relaxation

Polymer and resonance frequency	Measure of crosslinking	E_a/kJ mole^{-1} Low temperature process	High temperature process	
Cured natural rubber[a] 20 MHz	3.5[g]		30.8	
	10.5		23.8	
	20.0		21.5	
Polyethylene[b] 90 MHz	0[h]	8.4[l]	7.5[m]	—[n]
	0.5	7.4	7.3	18.8
	1.0	7.0	6.7	20.1
	1.5	8.1	7.1	22.6
	2.0	9 2	6.8	24.5
	2.5	10.0	6.4	26.4
Polypropylene[c] 90 MHz	0[h]	7.1[l]	16.1	
	0.5	7.1	15.0	
	1.0	7.1	13.6	
	1.5	6.8	13.1	
	2.0	7.4	11.7	
	3.0	7.5	10.3	
Poly(methyl acrylate)[d] 20 MHz	0[i]		50.2	
	1.8		43.9	
	4.7		40.5	
	9.4		41.0	
Polybutadiene[e]	0[j]		66.1	
	140		86.6	
Polyurethane[f] based on THF 21 MHz	0.70[k]		20.5	
	2.18		24.3	
	2.45		20.1	
	3.30		19.7	
	4.70		20.9	

[a] Taken from Ref. [19]; [b] taken from Ref. [69]; [c] taken from Ref. [68]; [d] taken from Ref. [67]; [e] taken from Ref. [73]; [f] taken from Ref. [70]; [g] phr of sulfur; [h] % of dicumyl peroxide; [i] number of crosslinking units per weight average PMA chain; [j] average number of repeat units between crosslinks; [k] concentration of chains between network joints, (mole/cm^3) × 10^4; [l] methyl group rotation; [m] segmental motion; [n] motions within the crystalline part of the polymer

Higher values of $T_{1\,min}$ and lower T_1 values above the minimum, are found to be proportional to the crosslink density in crosslinked polydimethylsiloxane networks (PDMS) (Fig. 7) [72]. The position of $T_{1\,min}$, which depends on frequencies of segmental motion, is very slightly influenced by the degree of crosslinking for a lightly crosslinked network (more than 10 DMS units between network points), but smaller spacings between network points give rise to a significant change in segmental motions. The same behavior for $T_{1\,min}$ and high temperature T_1 values is observed in crosslinked *cis*-1,4-polybutadiene (network chain density ranking from $6 \cdot 10^3$ to $3.3 \cdot 10^6$ mol/cm^3) [54]. However, in the same polymer T_1 values at temperatures below the T_1 minima are found to be nearly insensitive to the degree of crosslinking, whereas

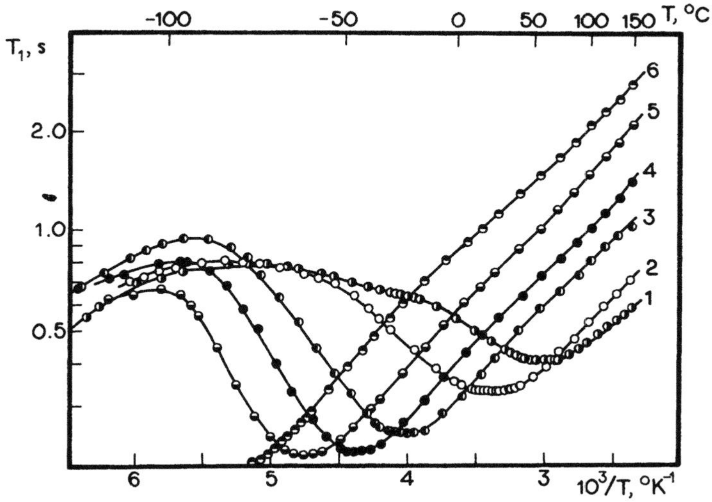

Fig. 7. Temperature dependence of T_1 at a proton frequency of 88 MHz for tetrafunctional network PDMS specimens with the following numbers of dimethylsiloxane units between network points: **2** (1), **3** (2), **6** (3), **9** (4), **30** (5) and for the linear PDMS (6) (reprinted from Ref. [72])

above the minima the effect of crosslinking is observable [73]. The latter results indicate that the high-frequency, short-range motions of segments between crosslinks are not affected by the motional restrictions imposed by the crosslinks. Only a small dependence of T_1 on radiation dose was observed in crosslinked polydimethylsiloxanes [74].

Spin-lattice relaxation times for crosslinked polystyrene (with a content of crosslinking agent from 0.5–2 mole %), swollen in toluene (polymer concentrations 20 and 30 mole %) show a shift of the V-curve to higher temperatures with increasing crosslink density and decreasing polymer concentration. Consequently, lower T_1 values with increasing crosslinking are observed in the high temperature region, and higher values in the low temperature region. However, with a 40 mole % concentration of polystyrene there is little effect of crosslinking on the T_1 above the minima, while in the low temperature region relaxation times are almost independent of the degree of crosslinking. It should be noticed that the formation of nonchemical fluctuation networks (based on hydrogen bonds) has a similar effect on T_1; in this case the T_1 minima are shifted to higher temperatures with the increasing number of hydrogen bonds, but owing to the lack of change in the T_1 values in the high temperature region, the V-curve becomes narrower [75].

Side group relaxation in individual crosslinked polymer systems could exhibit different behavior. i) The T_1 minimum due to the methyl group reorientation in diglycidyl ether of bisphenol-A (DGEBA) cured with 4,4′-methylenedianiline (MDA) is broadened and occurs at lower temperature as compared to the uncured resin. This relatively small effect implies a slightly higher E_a (23.0 and 19.7 kJ mole^{-1} for the cured and uncured resin, respectively) [53,76]. ii) The lowest temperature T_1 minimum in irradiated PVC-TEGDMA [poly(vinyl chloride) — tetraethylene glycol dimethacrylate] systems near the gel point are found to be shifted 5 degrees towards higher temperatures, indicating the restriction of the methyl group rotation due to the introduction of crosslinks [77]. iii) The position of the methyl group T_1 minima in polypropylene and low-pressure polyethylene are not affected by the degree of crosslinking, indicating that network formation does not influence the molecular dynamics of the CH_3 groups [68,69]. From the T_1 vs. temperature behavior it was suggested that the T_1 is closely related to the free volume in the linear polymer and network, respectively [78,79]. Furthermore, it was observed that the values of T_1 at the curve maximum in crosslinked polypropylenes (298 K) exhibit a steady drop with increasing degree of crosslinking [78]. More accurate measurements reveal two transitions at 293 and 313 K. Both of them change with the degree of crosslinking [68].

Activation energies for segmental motions in polymer networks, as calculated from slopes of log T_1 vs. $1/T$ plots and assuming a single correlation time, are generally lower than E_a in uncrosslinked polymers. This behavior is due to the lower values of T_1 above $T_{1\,min}$ in networks and, consequently, a smaller slope of the V-curve which is the most influential parameter in the determination of E_a. The use of different distribution functions $G(\tau_c)$ influences the absolute values of E_a but the general relationship with increasing crosslinking usually remains the same. However, a difference in one order of magnitude is found in the activation energies for PDMS networks if they are calculated from the diagrams of correlation frequencies (ν_c) and slopes of the V-curve, respectively. With the assumption that the position of the T_1 minimum is to a lesser extent dependent on the distribution of τ_c, E_a calculated from ν_c diagrams (based on the T_1 minimum and T_2 relaxation) for PDMS networks, are in the range 71–88 kJ/mole (NMR frequency 88 MHz); simultaneously, values of 7.5 to 9.2 and 10.5 to 12.1 kJ/mole were obtained from low and high temperature slopes of the V-curve, respectively [72].

Spin-lattice relaxation measurements of unsaturated polyesters based on maleic anhydride, phthalic anhydride and propylene glycol (and styrene as a second component) show two components, indicating the presence of fast and slow relaxing protons in the system. The fast component disappears during the crosslinking and the T_1 of the slow relaxing component changes with the curing time. The fractional loss of mobility is a measure of the degree of curing and can be determined as the ratio of correlation frequencies (ν_c) for the general molecular motions between particular curing times and the final limiting value, at a specified temperature. The energy of activation for the crosslinking process, calculated from a semiempirical expression for the time dependence of the fractional loss of mobility and the Arrhenius plot of the kinetic constants, is found to be 56.8 kJ per mole of double bonds [80].

A survey of calculated E_a for relaxation processes in crosslinked polymers calculated from T_1 measurements assuming a single τ_c is given in Table 2.

3.2.2 Spin-Lattice Relaxation in the Rotating Frame

Relaxation in the rotating frame is determined by the strength of the spin-locking field B_1. Therefore, $T_{1\varrho}$ times are sensitive to the intermediate frequencies $\omega_1 = \gamma B_1$, in the kHz region [49]:

$$\frac{1}{T_{1\varrho}} = C\left[\frac{3}{2}\,\frac{\tau_C}{1+4\omega_1^2\tau_C^2} + \frac{5}{2}\,\frac{\tau_C}{1+\omega_0^2\tau_C^2} + \frac{\tau_C}{1+4\omega_0^2\tau_C^2}\right] \tag{19}$$

Thus, larger-scale cooperative chain motions can be observed using this technique. A $T_{1\varrho}$ vs. temperature plot is generally analogous to a T_1 curve with the lower time values and minima shifted to lower temperatures (Fig. 4). In the case of rapid motions ($\omega^2\tau_c^2 \ll 1$), $T_{1\varrho}$ is identical to T_1 and, assuming a single correlation time, both relaxation times are equal to $1/5\ C\tau_c$ (Eqs. (15) and (19)).

The pulse sequence which provides measurement of the the $T_{1\varrho}$ relaxation time is shown in Fig. 8. The second 90° pulse following the initial 90° pulse causes the phase shifting and the magnetization is spin-locked along the $\mathbf{B}_1$ field. Since $B_1 \ll B_0$, the procession frequencies are usually three orders of magnitude smaller than those in the static magnetic field. By the turning off of the B_1 field after a suitable spin-lock period the FID in the rotating frame is observed. The corresponding $T_{1\varrho}$ times are obtained from the plot of the FID intensities vs. the length of the spin-locking pulse

It is found that $T_{1\varrho}$ values in crosslinked polybutadiene at the temperatures below the minimum, as in the case of T_1, are nearly insensitive to crosslink density; only highly crosslinked networks exhibit significantly lower $T_{1\varrho}$s [73]. However, the temperature dependence of $T_{1\varrho}$ for the polybutadiene networks (prepared by peroxide crosslinking in both, solution and bulk) show that bulk crosslinked networks relax faster, probably due to the effect of entanglements trapped within the network during crosslinking, which act as effective crosslinks, and decrease the relaxation times [81].

The $T_{1\varrho}$ in epoxy resins DGEBA-MDA cured at 54, 100 and 180 °C, show a greater sensitivity to the segmental motion as compared to T_1 (Fig. 9) [53,76]. Two $T_{1\varrho}$ minima in the monomer DGEBA correspond to methyl group reorientation (−117 °C) and general molecular motion (8 °C), respectively. The almost identical positions of the methyl minima in cured resins indicate that the activation energy for methyl reorientation is the same for uncured and cured resin, or possibly slightly higher for resins cured at 100 and 180 °C. However, a change in segmental motion has a significant effect on the temperature dependence of $T_{1\varrho}$. The almost temperature-independent $T_{1\varrho}$ above −50 °C in the resin cured at 54 °C suggests that there are no clearly defined segmental motions, while in resins cured at higher temperatures two minima appear at 26 °C and 90 °C, respectively. The decreasing intensity of the lower temperature minima (faster motions) and increasing intensity of higher minima (slower motions), in the sample postcured at 180 °C are probably due to the additional crosslinking, since the resin cured at 100 °C is expected to have an extent of reaction greater than 90%.

In some crosslinked polymer systems, like air-cured alkyd polyester paints, two separate $T_{1\varrho}$ decay processes are observed. Longer $T_{1\varrho}$s in these systems is associated with a more rigid environment for the spins; the fractions of the two $T_{1\varrho}$ populations indicate the amount of rigid and mobile protons, respectively [82].

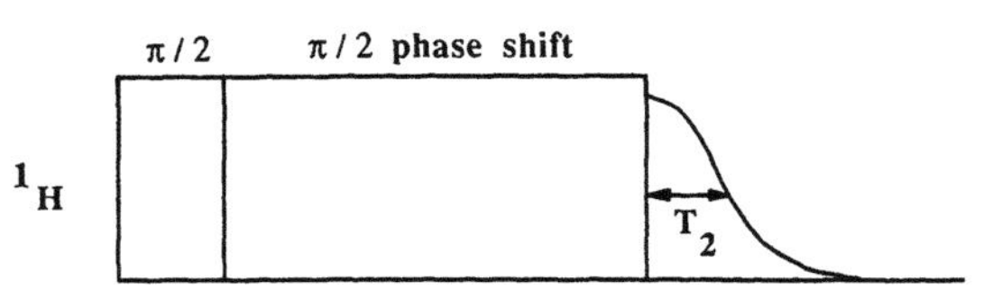

Fig. 8. $^1H\ T_{1\varrho}$ pulse sequence

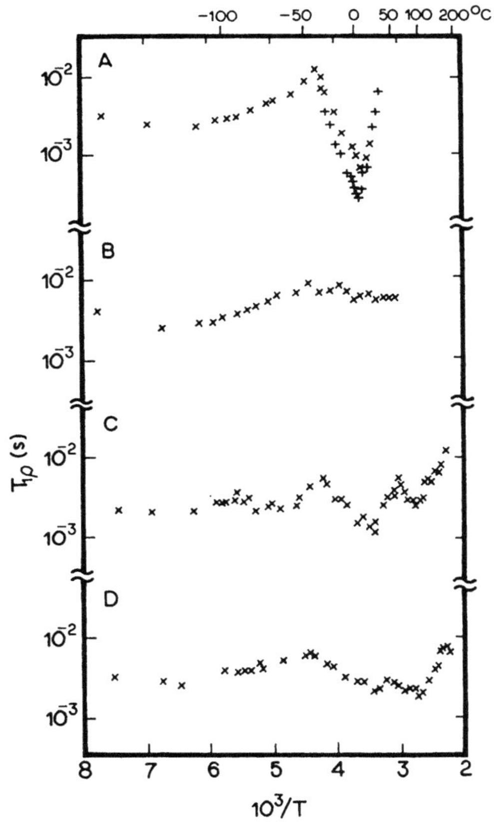

Fig. 9. Plot of log $T_{1\varrho}$ relaxation time vs. reciprocal temperature for the diglycidyl ether of bisphenol-A: uncured (A) and cured with methylenedianiline at 54 °C (B), 100 °C (C) and 180 °C (D) (adapted from Refs. [53] and [76])

3.2.3 Spin-Spin Relaxation

The spin-spin relaxation time can, in principle, be measured from the FID following the 90° pulse (one-pulse experiment; Fig. 10a). However, the application of this simple experiment is limited only to very short FIDs, and consequently short T_2s, because of the inhomogenities in the laboratory field. The line width in magnetic resonance is proportional to T_2, but the observed linewidth, T_2^*, has also a contribution from the magnetic field term:

$$\frac{1}{T_2^*} = \frac{1}{T_2} + \frac{\gamma \, \Delta B_0}{2} \tag{20}$$

For example, the contribution of the magnet inhomogeneity for the magnetic inductions of 1.4 T (60 MHz) to the observed T_2^* becomes significant at T_2 values of about 0.5 ms. In order to overcome the problems caused by the magnet inhomogeneity (generally for T_2 values in the millisecond range), spin-echo techniques are applied [83]. One of the commonly used pulse sequences is the Carr-Purcell sequence (Fig. 10b), which allows the observation of a train of n echoes obtained by a single

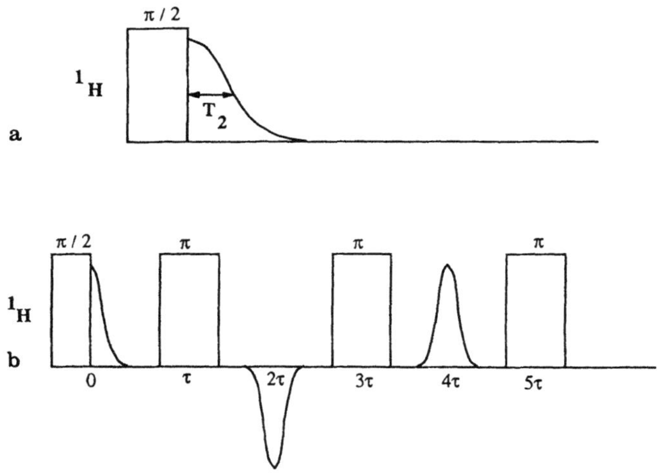

Fig. 10a and b. ^{1}H T_2 pulse sequence: free induction decay following a 90° pulse (**a**); Carr-Purcell pulse sequence (**b**)

pulse sequence; from the intensities of echoes, the T_2 time is calculated. The main limitation of the spin-echo techniques is their sensitivity to the diffusion effect, which is dependent on the spatial magnetic field gradients (G), the diffusion coefficient (D) and the time during which the diffusion occurs (τ). The echo intensities I at the time t can be expressed as:

$$I(t) \propto \exp\left[-\left(\frac{t}{T_2}\right) - \frac{1}{3}\gamma^2 G^2 D\tau^2 t\right] \tag{21}$$

By chosing short values of τ, the effect of diffusion may be eliminated, because the diffusion is effective in reducing the echo intensity only during the 2τ period. However, there is a modified version of spin-echo pulse sequences which provides the measurement of the spin diffusion coefficients (D) by NMR, as it will be shown later.

Spin-spin relaxation times (T_2) in polymer systems range from about 10^{-5} s for the rigid lattice (glassy polymers) to a value greater than 10^{-3} s for the rubbery or viscoelastic state. In the temperature region below the glass transition, T_2 is temperature independent and not sensitive to the motional processes, because of the static dipolar interactions. The temperature dependence of T_2 above T_g and its sensitivity to low-frequency motions, which are strongly affected by the network formation, make spin-spin relaxation studies suitable for polymer network studies.

The free induction decay following 90° pulse has a line shape which generally follows the Weibull functions (Eq. (22)). In the homogeneous sample the FID is described by a single Weibull function, usually exponential (Lorentzian) ($\mu = 1$) or Gaussian ($\mu = 2$). The FID of heterogeneous systems, such as highly viscous and crosslinked polydimethylsiloxanes (PDMS) [84], hardened unsaturated polyesters [85], and compatible crosslinked epoxy-rubber systems [52] are actually a sum of three decays. However, most of the crosslinked polymers exhibit FIDs which consist of two components. For example the FIDs of PDMS irradiated with doses sufficient to cause gelation exhibit two exponential decays (Fig. 11) [86]. The double exponential expression can be written in the more general form as a sum of two Weibull functions:

$$\frac{I(t)}{I(0)} = f \exp\left[-\left(\frac{t}{T_{2S}}\right)^{\mu_1}\right] + (1-f)\exp\left[-\left(\frac{t}{T_{2L}}\right)^{\mu_2}\right] \quad (22)$$

where I(0) and I(t) are signal intensities at time zero and t, respectively; T_{2S} and T_{2L} are the long and short transverse relaxation times, respectively, and f is the fraction of the total number of rigid protons having a shorter T_2 component. In the case of exponential decay both exponents μ are equal to 1.

Generally, T_2 relaxation times are very sensitive to slower relative translational motions of the polymer chains and can provide information on intramolecular couplings, such as chemical crosslinks and chain entanglements. Numerous studies on both permanent and temporary networks are presented in a series of papers by Charlesby and co-workers [74,86–94]. In the case of extracted polymer networks, T_{2S} relaxation is observed in the crosslinked (gel) fraction, while T_2 relaxation occurs in the soluble fraction of the irradiated polymer [86]. It is shown that the fraction of more mobile protons, (1-f), has the same general trend with increasing

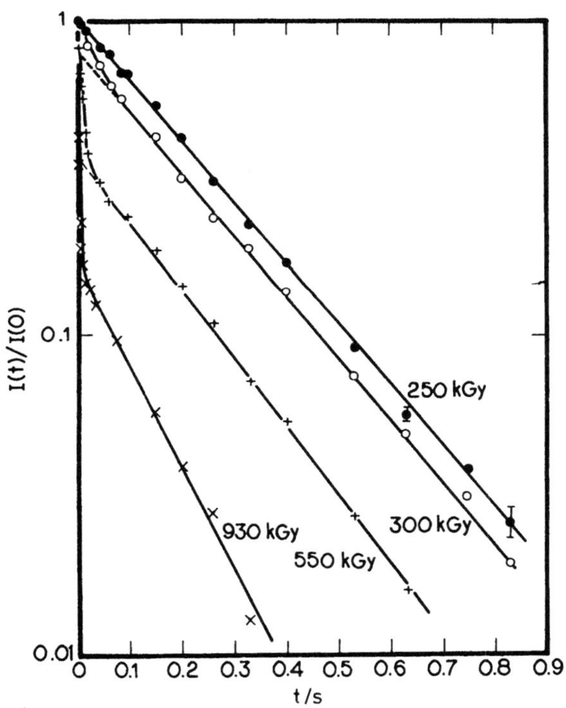

Fig. 11. Experimental transverse relaxation plots for PDMS at four different doses showing the double exponential behavior at doses of 300 kGy and above (adapted from Ref. [86])

dose as the sol fraction. Although the NMR measurements give a slightly lower (5–10%) sol content as compared to the extraction method, the NMR data fit very well to the Charlesby-Pinner equation over a wide range even close to the gel point. The T_{2L} component in the sol fraction at 10 °C is found to be related to the number-average molecular mass M_n, while in the separated gel fraction T_{2S} is directly proportional to the average molecular mass between crosslinks (or entanglements) M_c [74]:

$$T_{2S} = 6 \cdot 10^{-7} M_c \tag{23}$$

The studies of model SBS systems show that the T_{2L} component (corresponding to the PB chain between PS blocks) increases proportionally to the number of mobile protons until the critical molecular weight for the formation of entanglement of the rubber segment (40000) is reached; then the two T_2s are observed for a linear PB [95]. Calculated and NMR determined mobile proton fractions are in a good agreement, except for the smallest (1-f) values.

However, some other uncrosslinked completely soluble polymers of high molecular weight also show two component decays. Examples are PDMS (M_w = 260000) [90] and cis-polyisoprene (M_n = 150000) [92]. Since there is no chemical network present, the complex FID for these polymers is explained in terms of the entanglements between polymer chains which cause fluctuating, temporary networks, lasting long enough to impose motional restrictions during the NMR experiment [90]. The investigation of dynamic network formation in polystyrene shows that a single exponential FID is obtained with PS M = 3000 and M = 660000, respectively. In the former, the entanglement density is insufficient to form any network (f = 0), while in the latter a single FID corresponds to a fully entangled network (f = 1). However, PS with M = 59000 exhibits both components T_{2L} and T_{2S} [94]. Comparison with the solubility data shows that the entanglements behave as the crosslinks from the NMR point of view. Moreover, the average molecular weight per entanglement point as calculated from NMR is in good agreement with the results obtained from the M_w dependence of the bulk viscosity [87].

Uncrosslinked polyurethanes (PU) consist of two domains, i.e. hard and soft segments, which also give rise to a two component FID in which the fast Gaussian component corresponds to the hard, and the exponential component to the soft segments [96]. Since a very broad distribution of correlation times also can cause a two-component FID, the Goldman-Shen pulse sequence [97] (in which the second −90° pulse is applied at the particular time, when T_{2S} component is dephased completely and T_2 component is not dephased significantly), is applied in order to find the origin of the two T_2 relaxation times. The slow exchange of magnetization observed in PU samples indicates that the spins producing the two different components exist in two different phases, physically separated by large distances. Crosslinking of polyestsr urethanes with peroxide causes a less rapid decay of the fast component (T_{2S}) and a more rapid decay of the slow component (T_{2L}), as a result of the increased mobility of the hard phase, and a decreased mobility of the soft phase [98]. The behavior of particular T_2 components with increasing crosslinking (% of peroxide), their ratio (T_2/T_{2S}) as a measure of the differences in mobilities of

Table 3. Spin-spin relaxation parameters for crosslinked polyether urethanes[a]

% of peroxide	$T_{2S/\mu s}$	$T_{2L/\mu s}$	T_{2L}/T_{2S}	f
0.0	14.7	218.0	14.8	0.262
0.5	16.2	187.0	11.5	0.293
1.0	20.2	183.0	9.1	0.274
2.0	32.6	175.0	5.4	0.286

[a] Reprinted from Ref. [97]

the two phases and the fraction of protons associated with the hard phase (f) at 28 °C, are given in Table 3.

From the molecular mobilities of the soft and hard phases, it appears that the increase in phase mixing caused by crosslinking, is more important in determining phase mobility than a simple increase in crosslink density. The fraction of protons in the hard phase (f) is relatively independent of crosslink density at 28 °C. However, f in linear and crosslinked PEU have different temperature behaviors (Fig. 12). Incurred PEU exhibits a single FID below —20 °C and again above 80 °C, when both phases are below and above T_g, respectively. In the region between approximately 10 and 75 °C, at which temperatures the hard and soft phase undergoes a transition, respectively, the fraction of protons in hard domains remains relatively constant. However, f in the crosslinked PEU decreases in a continuous fashion over a relatively

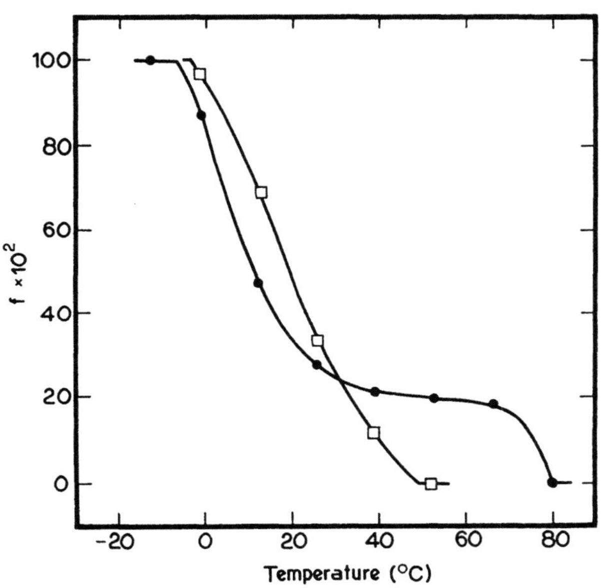

Fig. 12. Fraction of protons (f) attributed to the hard phase for the linear polyurethane (●) and polyurethane crosslinked with 2% of peroxide (□) (adapted from Ref. [98])

small temperature interval. The shift of the soft phase temperature transition to higher temperatures, and the reverse situation at higher temperatures, indicate that crosslinked PEU does not contain regions that are pure in hard-phase segments as does linear PEU, and that the composition of the crosslinked PEU varies in a continuous fashion from regions rich in hard segments to regions rich in soft segments [98].

The temperature dependences of the mobile proton fraction (1-f) for mixtures of poly(vinyl chloride) (PVC) and bifunctional tetraethylene glycol dimethacrylate (TEGDMA) monomer exhibit three minima in the investigated temperature range (Fig. 13). They are attributed to the following motional hindering changes: temporary ordering of some PVC molecules with a consequent loss of mobility (I); thermally activated TEGDMA polymerization (II), and the onset of crosslinking on PVC (III). It is found that T_g for crosslinked samples, as measured by a temperature at which T_{2L} appears, increases almost linearly with irradiation doses up to 1 kGy. However, it should be noted that significant changes are produced by heat treatment at 95 °C,

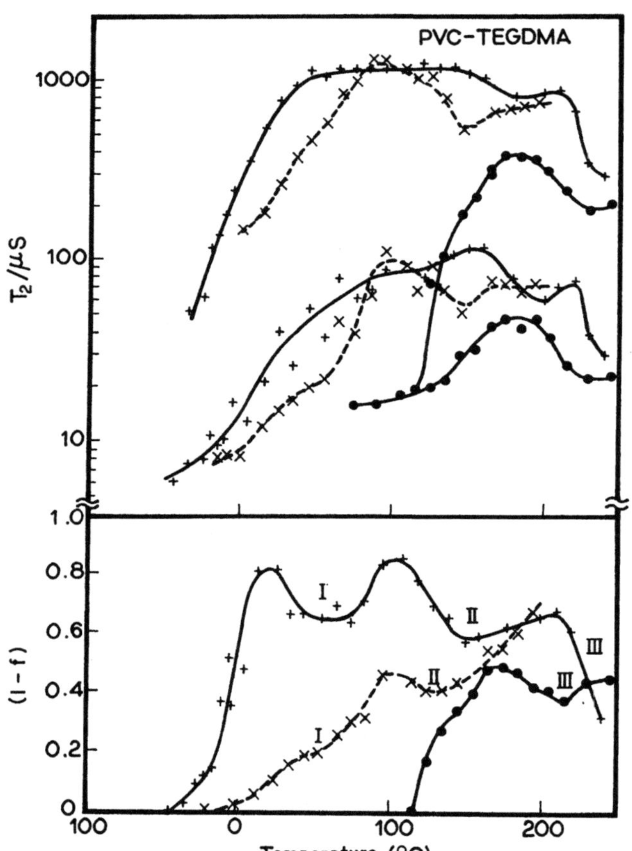

Fig. 13. T_2 and the mobile proton fraction (1–f) vs temperature for 0.0 (+), 0.6 (×) and 50 kGy (○) PVC-TEGDMA mixtures (adapted from Ref. [77])

which include the disappearance of the minimum I, and the shifting of the T_2 transition to higher temperatures [77].

The existence of two T_2 relaxation times was also observed in the radiation cured photopolymers [99], and studies of tri- and tetra-functional network organosilicon polymers with rings at the network points [100]. The effects of crystallization of polydimethylcarbosiloxane networks (PDMCS) [101], and water sorption and stoichiometrical composition of the cured DGEBA/DETA resins on phase structure and mobility [102] were studied.

The two component FID (and corresponding superposed broad line spectra), for some polydimethylsiloxanes (PDMS) arises from crystalline and amorphous components. Polymers highly crosslinked with dicumyl peroxide exhibit quite similar data, but after the removal of all unlinked material, the FID consists only of a single component, indicating an almost complete reduction in crystallinity due to the crosslinking [84].

Crosslinked polyurethanes based on THF copolymers exhibit different FID shapes (single Lorentzian, single Gaussian or sum of Lorentzians), depending on the network structure [70,71]. In polyether urethanes (PEU) with the T_2s differing less than a factor of two, a single T_2 relaxation time for nonexponential FIDs was calculated. By determining τ_c from T_2^{calc} and ΔM_2 of the rigid lattice, the correlation times are found to be slightly shorter for a linear PEU than for a crosslinked one (crosslink density $= 1.46 \times 10^{-4}$ mole/cm^3). Likewise, a deviation in the temperature dependence of E_a from the Arrhenius law (inflection) is ascribed to the presence of two τ_cs for rapid and slow segments, respectively. However both linear and crosslinked PEU, have the same apparent activation energies (45.2 kJ/mole) [103].

Spin-spin relaxation and broad line NMR studies of the curing of unsaturated polyesters with styrene, reveal the existence of three T_2 relaxation times, although the shortest T_2 is not directly observed. However, from the loss of initial magnetization, I(0), during the crosslinking process the fraction of the three types of protons could be estimated (Fig. 14): I) f_1 and the corresponding T_{21} time arising from very mobile protons are ascribed to unpolymerized styrene; II) f_2 and the corresponding T_{22} time arising from less mobile protons are ascribed to free polyester molecules; III) the third fraction f_3, developed in the course of the curing, is determined from the difference in I(0), i.e. $f_3 = 1 - (f_1 + f_2)$. The corresponding T_{23}, as determined from broad line NMR, is almost temperature independent (in the range 10^{-5} s) and it is ascribed to the immobile protons in the cured glasslike regions [85].

A comparison with broad line measurements shows that the fraction f_3 corresponds to the fraction of the broad Gaussian area (immobile protons), while both fractions f_1 and f_2 correspond to the narrow Lorentzian area (mobile protons), in the two component broad line spectra. The changes of the particular T_2s and proton fractions during the curing are shown in Fig. 14. It is found that the resins cured at room temperature still have about 7 to 18% of mobile protons (styrene); resins cured at increasing temperatures have a smaller content of mobile protons (3 to 7%). Assuming a Gaussian distribution of the mean number of jump times τ for which a molecule remains in the mobile phase, an average radius of the mobile phase can be determined from the T_{21} data. The radius is decreasing with the curing time from 30 to 0.9 nm (for curing times from 20 to 300 minutes). Based on NMR and light scattering data, a model for curing of unsaturated polyesters is proposed (Fig. 15).

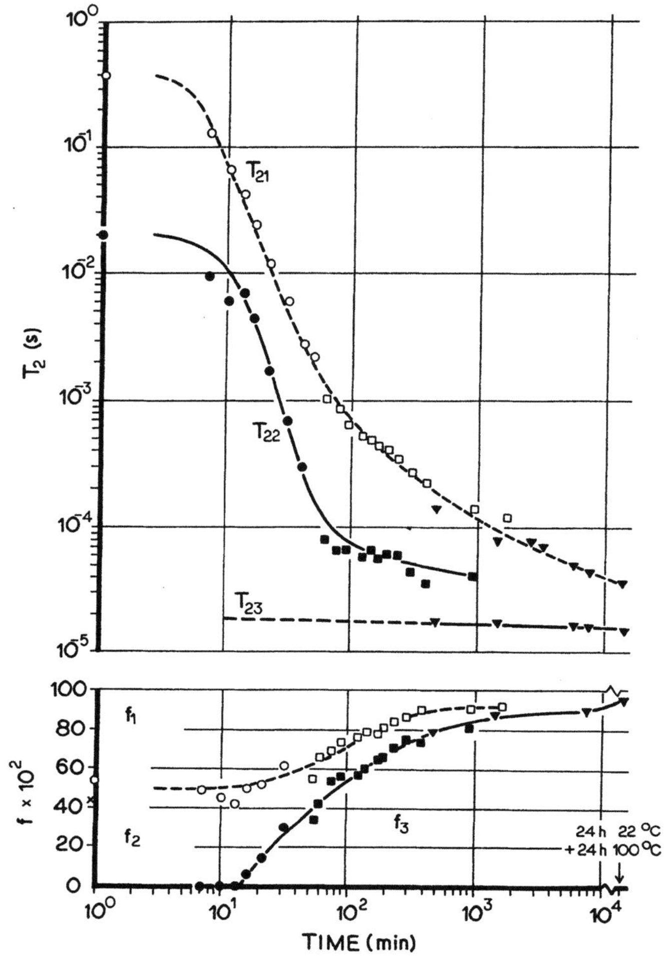

Fig. 14. Dependence of the relaxation times T_{2i} and the fractions of protons with different mobility (f_i) for unsaturated polyester on the curing time, as measured from broad line NMR (▼), Hahn spin-echo (□) and Carr-Purcell pulse sequence (○). Symbol × indicates the initial distribution of styrene and unsaturated polyester protons (adapted from Ref. [85])

According to this model the structure of the crosslinked polymer is not considered as a homogeneous infinite network of polyester molecules, connected by styrene bridges. Since the styrene is a poor solvent for polyester chains the copolymerization takes place at statistically placed radicals ($R^{\cdot}$) [85].

The multicomponent FID observed in the amine-hardened diglycidyl ether of diphenylpropane is believed to be a consequence of nonuniformity in the kinetic behaviour of the various network fragments; in the case of three components, protons are ascribed to the following groups: "ends" (molecules attached with with one end to a nodal fragment), spacers (chains between crosslinks) and nodal fragments, respectively [104].

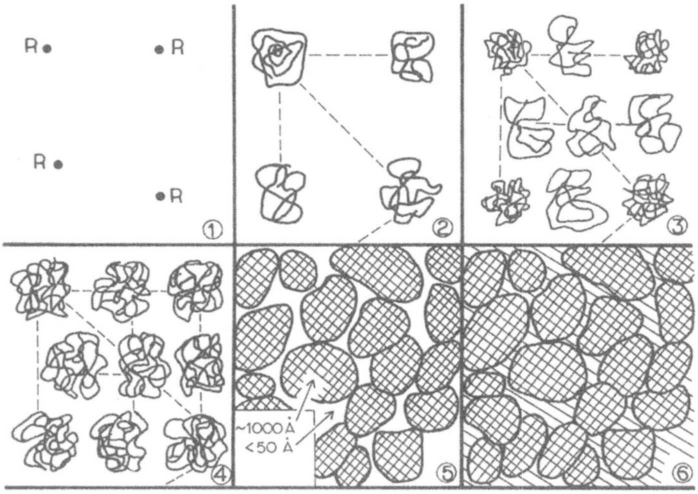

Fig. 15. Model for the crosslinking of unsaturated polyester resins (reprinted from Ref. [85])

Constant restrictions on segmental motion (incomplete averaging of the dipolar interactions), imposed by crosslinking, cause a T_2 plateau in the high temperature region in the spin-spin relaxation time vs temperature dependence. Swollen PS crosslinked with 1,6-bis-(*p*-vinylphenyl)hexane (with mole% of crosslinking agent 0.5 to 2) [105], as well as solid crosslinked *cis*-1,4-polybutadienes [73] exhibit a single T_2 relaxation time accompanied with decreasing T_2 values of plateaus as the crosslinking density increases (Fig. 16).

Line shape analysis are of considerable importance in the relaxation studies of linear and crosslinked polymers, particularly of elastomers. It is observed that in the vicinity of the glass transition the abrupt change in polybutadiene T_2 is accompanied by changes in proton pulsed FID line shapes, which can be described with the Weibull functions (Eq. (22) [106]. The exponent μ decreases with increasing temperature, from about 2.8 at -150 °C, and the FID, passing through the Gaussian ($\mu = 2$), reaches the Lorentzian exponential line shape ($\mu = 1$) at T_g, while in the rubber like state, μ has a value of about 1.4. With increasing crosslink density in natural rubber the minimum in the temperature dependence of μ is shifted to higher temperatures and the line shapes of FID are changed from Gaussian to Lorentzian in the temperature region above the glass transition [107].

A pseudo solid-like behavior of the T_2 relaxation is also observed in i) high M_n fractionated linear polydimethylsiloxanes (PDMS), ii) crosslinked PDMS networks, with a single FID and the line shape follows the Weibull function ($\mu = 1.5$) [88] and iii) in uncrosslinked *cis*-polyisoprenes with $M_n > 30000$, when the presence of entanglements produces a transient network structure. Irradiation crosslinking of polyisoprenes having smaller M_n leads to a similar effect [91]. The non-Lorentzian free-induction decay can be a consequence of a) anisotropic molecular motion or b) residual dipolar interactions in the viscoelastic state.

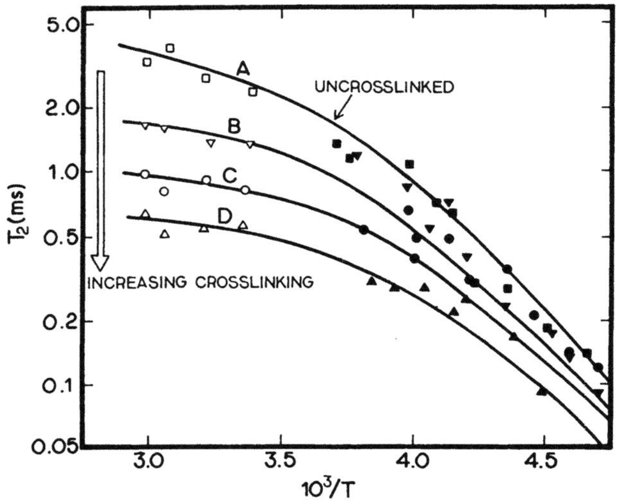

Fig. 16. Temperature dependence of T_2 at 24.3 MHz for *cis*-polybutadiene; linear PB (A) and samples with the following average number of repeat units between crosslinks: 140 (B), 44 (C) and 14 (D). Open symbols denote T_2 measurements obtained by spin-echo technique (reprinted from Ref. [73]) with permission)

The dependence of T_2 values on the molecular parameters is explained by a simple model for the chain with both ends fixed, and defined by the number of statistical segments, Z, in the chain between the crosslinks [108]):

$$Z = 1/ns \tag{24}$$

where n is the mole fraction of monomers bearing crosslinks and s is the number of monomers in one segment (thermodynamic rigidity). Assuming that the constant value of T_2 in the plateau region is a consequence of non-averaged dipole-dipole interactions (which give rise to the residual part of the second moment), and Lorentzian line shapes, the following equation is established:

$$T_2 \simeq \frac{5.6z}{\gamma(\Delta M_2)^{1/2}} \tag{25}$$

where ΔM_2 is the rigid-lattice second moment. The calculated T_2 values are of the same order as experimental, but are substantially smaller as the degree of crosslinking becomes very small. This is probably due to the increased content of loose ends in the network.

The dependence of T_2 values in the plateau region on the crosslink density, indicate the effect of restricted network dynamics on the low-frequency, long-range configurational motions. A plot of normalized T_2 for crosslinked polybutadienes (PB), (ratio of T_2 for crosslinked PB over T_2 for uncrosslinked PB), is shown in Fig. 17 [73]).

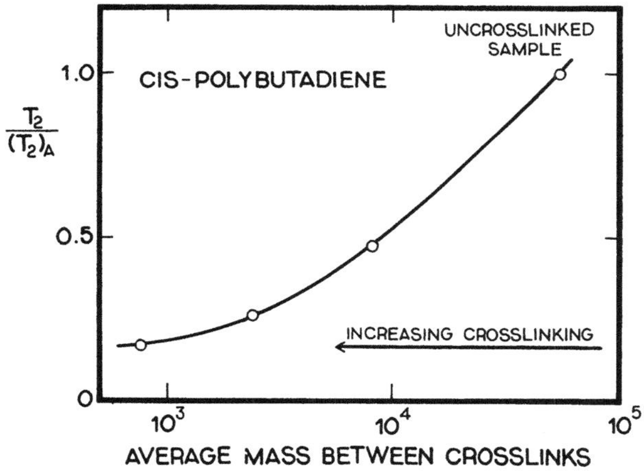

Fig. 17. Relationship between T_2 values measured at 60 °C and 24.3 MHz and the average mass between crosslinks. T_2 values are normalized to the T_2 value found for uncrosslinked PB sample (reprinted from Ref. [73]) with permission)

Constraints upon the molecular mobility also exist in the fluctuation networks based on strong hydrogen bonds. Copolymers of styrene, acrylic acid and methacrylic acid in toluene solution, show a shift in the abrupt change of T_2 to higher temperatures with increasing concentration of carboxyl groups (i.e. hydrogen bonds), accompanied by the increasing apparent activation energy calculated from the low temperature T_2 slopes:

$$E_a = \frac{-Rd(\ln T_2)}{d(1/T)} \tag{26}$$

A significant difference between chemically crosslinked networks and hydrogen bonding (fluctuation) networks, is the absence of the plateau region in the high temperature region.

Uncrosslinked poorly compatible epoxy-rubbers exhibit two T_2 times in the temperatures above −50 °C. The T_{2S} component ($f = 0.05$) has a constant value (12 µs) up to 27 °C; above this temperature it is no longer observable. This component is ascribed to ordered regions, probably including urethane end groups, which melt at temperatures above 27 °C. In the crosslinked samples, an ordered phase (T_{2S}) is preserved up to about 80 °C, due to the stabilizing action of chemical crosslinks; the T_{2L} component reaches a constant value of the plateau which is found to be independent of the concentration of the rubber component. The independence of the T_{2L} plateau values on the rubber component concentration, as well as measurements of the cross nuclear magnetic relaxation times, indicate that a poorly compatible urethane rubber and epoxy oligomer form an incompatible network with no interaction between the components of the system. However, compatible epoxy-rubber systems exhibit three T_2 times: a) a T_{21}, ascribed to the motion of low molecular weight residues of the rubber molecules which have a plasticizing effect on the network; b) a T_{22} arising from motions of "incorporated" fragments of the

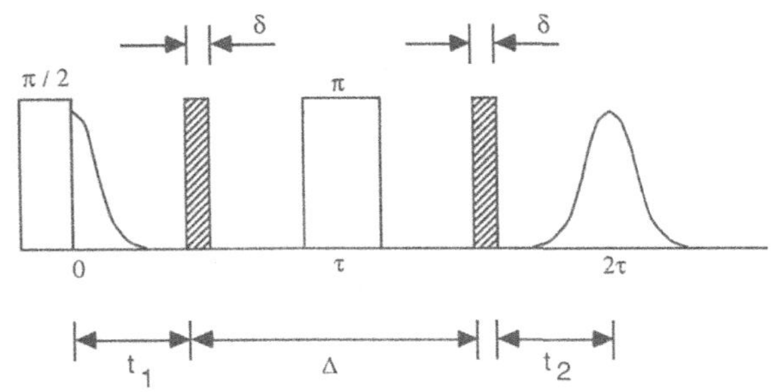

Fig. 18. Pulsed-gradient spin-echo (PGSE) pulse sequence

rubber molecules and terminal fragments, and c) a T_{23} connected with the crosslinked regions. Compatible components have a marked influence on the dynamics of segmental motions [52].

Spin-spin relaxation dynamics were also used in the study of the kinetics of the vulcanization of polysulphide rubbers. The T_2 values decrease with the course of the reaction and the time dependence of log (T_2/T_2^0), where T_2^0 corresponds to the time equal to zero, exhibits an inflection. The inflection point is attributed to gel formation, and the reaction rate constants for the two separate processes are determined from the T_2 data. It was also observed that the addition of carbon black reduces T_2 by a factor of 2 or 3, because vulcanization occurs both through the thiol groups and by the chemical reaction between the polymer and carbon black [37].

Important information about molecular dynamics can be obtained from diffusion measurements. Although the spin-echo techniques are generally suitable for detecting diffusion effects (Eq. (21)), there are experimental limitations, mainly arising from the magnetic field gradient (G), which must be present all the times. In order to measure the diffusion coefficient (D), the pulsed-gradient spin-echo (PGSE) technique was developed [109]. The PGSE pulse sequence (Fig. 18) is based on a modified $90° - \tau - 180°$ spin-echo pulse sequence: the magnetic field gradient (G) is turned on at time t_1, and again later at the a time $t_1 + \Delta$. The gradient causes the dephasing of the magnetization. If the nuclei do not diffuse the magnetization will refocus with no loss in intensity after the 180° pulse (reverse the direction of the dephasing) and the second gradient pulse (produces an equal and opposite rephasing of a magnetization). However, a loss in the intensity will be observed if there is diffusion due to the partial refocusing.

The PGSE experiment can detect the presence of two or more diffusivities in the species at resonance. The intensity of the spin echo is a function of the duration δ, magnitude G and the time Δ between the applied magnetic field gradient pulses:

$$\frac{I(2\tau, x)}{I(2\tau, 0)} = \sum_{i=1}^{N} h_i \exp\left[-\gamma^2 D(M_i)\, x\right] \tag{27}$$

where

$$x = (G\delta)^2 \left(\Delta - \frac{\delta}{3}\right) \tag{28}$$

The additional correction terms involving the steady magnetic field gradient (G_0), which arises from the inhomogeneities in the laboratory magnetic field B_0 are relatively small compared to the pulsed field gradient G_0 and can be neglected to a first approximation. The term h_i is defined as:

$$h_i = kw(M_i) \exp\left[\frac{-2\tau}{T_2(M_i)}\right] \tag{29}$$

where k is chosen to normalize the sum of h_i to unity, and $w(M_i)$ is the weight fraction (normalized to unity) of molecular mass M_i.

The effect of curing on the diffusion of polymer and the curing agent is studied for the system of hydroxyl-terminated polybutadiene (R-45-M)/isophorone dissocyanate (IPDI). Both components contribute to the echo intensity and the plot of $\ln [I(\tau)/I(0)]$ vs $(G\delta)^2 (\Delta - \delta/3)$ consists of two exponentials (Eq. (22)); the fast component (the steep intial slope) is attributed to the IPDI, and the long component to the R-45-M. The dependence of both diffusion constants on the curing time is shown in Fig. 19. The accuracy for D_{fast} data is less pronounced than for the polymer $D(M_n)$, because only the first few data points are relevant for its determination. Furthermore, the low tail of the R-45-M molecular weight distribution nearly coin-

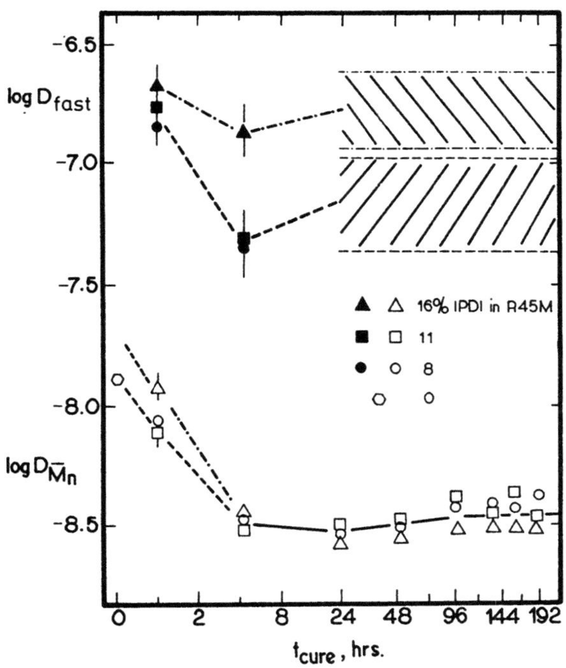

Fig. 19. Results for the 63 °C curing study for R-45-M containing 8% (□), 11% (○) and 16% (△) of curing agent IPDI. Diffusion of IPDI (top) and polymer molecules of weight M_n (bottom) is shown as function of curing time (abscissa scale is distorted). For IPDI diffusion after the first day, the upper error band corresponds to 16% concentration, the lower band to both 8 and 11% (reprinted from Ref. [110]) with permission)

cides with the molecular weight of IPDI (211 g/mol) resulting in a contribution of both species to the initial rapid echo attenuation, attributed only to the IPDI. (Thus, the D_{fast} are indicated in the form of error bands for the curing times longer than 24 h). The relatively small decrease of D for IPDI is in agreement with the insensitivity of the diffusion rate of small molecules to the crosslink density. However, the more precisely characterized $D(M_n)$ shows that most of the decrease takes place within the first 48 h, although at that stage the mechanical properties are not yet characteristic of a network. With longer curing times the values approach an asymptotic value, slightly higher than the minimum $D(M_n)$ observed after 24 h of curing. This slight minimum suggests the formation of aggregates large enough to move more slowly but not yet sufficiently constrained by the presence of a network. The polymer diffusivity in the early stages of curing is decreased by a factor of approximately three [110].

4 High Resolution NMR in Solutions

Rapid molecular motions in solutions average to zero the dipolar and quadrupolar Hamiltonian terms. Hence, weak interactions (chemical shift and electron-coupled spin-spin couplings) are the main contributions to the Zeeman term. The chemical shift term ($\mathbf{H}_\delta$) arises from the shielding effect of the fields produced by surrounding electrons on the nucleus:

$$\mathbf{H}_\delta = h\gamma \mathbf{I} \boldsymbol{\sigma} \mathbf{B} \tag{30}$$

where $\boldsymbol{\sigma}$ is the chemical shielding tensor (of second rank), described in terms of three diagonal values σ_{11}, σ_{22} and σ_{33} and the corresponding angles which determine the orientation of a principal axis system. Due to the rapid reorientations in solution, the diagonal tensor components are averaged to the isotropic chemical shift parameter $\sigma = 1/3\mathrm{Tr}\{\sigma_{ii}\}$.

The electron coupled spin-spin interaction between spin **I** and **S** is described by the term $\mathbf{H}_J$:

$$\mathbf{H}_J = h\mathbf{I}\mathbf{J}_{IS}\mathbf{S} \tag{31}$$

where $\mathbf{J}_{IS}$ is spin-spin coupling tensor which is averaged out in solutions to the scalar value of the spin-spin coupling constant, J_{IS}.

Thus, with the assumption of rapid isotropic bond reorientations, the Hamiltonian describing the two-spin system under high resolution conditions ($\mathbf{H}_{HR}$) consists of three terms, $\mathbf{H}_Z$, $\mathbf{H}_\delta$ and $\mathbf{H}_J$:

$$\mathbf{H}_{HR} = -h\gamma B_0(1 - \sigma_0)\,\mathbf{I}_z - h\gamma_S B_0(1 - \sigma_S)\,\mathbf{S}_z + hJ_{IS}\mathbf{I}\mathbf{S} \tag{32}$$

However, NMR spectra of swollen polymer gels under high resolution conditions usually exhibit line widths much larger than in the NMR spectra of analogous linear polymers, mainly owing to the insufficient averaging of dipolar interactions (residual H_D term). Anisotropy of magnetic susceptibility caused by the physical heterogeneity of the sample can also cause some line broadening [111].

4.1 High Resolution ^{1}H NMR

Line widths in ^{1}H NMR spectra of swollen gels are in the range between standard broad-line and high-resolution widths, depending on the degree of restricted molecular motions, i.e. degree of crosslinking. Usually the narrow lines are superimposed on a very broad background; integration of narrow lines for a crosslinked PMMA (with 0.5% of crosslinking agent) shows that only about 13% of the total nuclei contribute to the resolvable spectrum, which corresponds mainly to the linear PMMA and residual monomer; higher degrees of crosslinking (1% of crosslinking agent) causes a disappearance of narrow lines at room temperature (Fig. 20) [112].

Some swollen crosslinked polymer gels, such as styrene-divinylbenzene copolymers, exhibit high resolution spectra with so-called "super-Lorentzian" (SL) line shapes

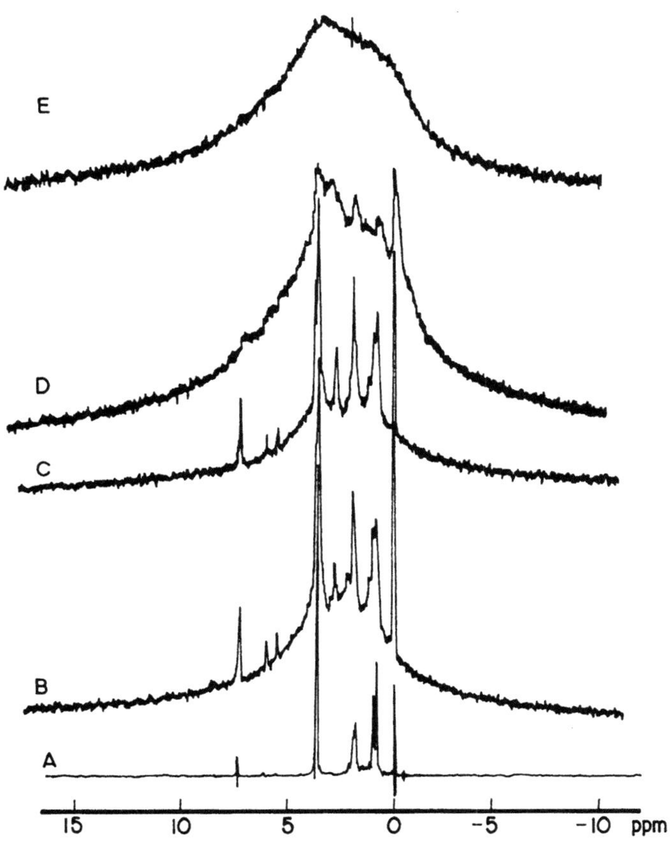

Fig. 20. Comparison of standard 100 MHz high resolution NMR spectra of linear and crosslinked PMMA (15% w/w in $CDCl_3$). Concentration of crosslinking agent (mol%): (A) 0; (B, C) 0.5; (D, E) 1. Temperature of measurement: (A, B, D) 60 °C; (C, E) 22 °C (reprinted from Ref. [112])

(Fig. 21), which could be in principle described by a certain distribution of τ_c (Eq. 17)[113]. Although the curves can be fitted with at least five different correlation times, the physical interpretation of the structural and dynamical phases in the system, corresponding to these observed τ_c is not clear[3]. Since SL line shapes disappear with spinning the sample at the magic angle, it is concluded that SL lines arise from insufficient averaging of the static intragroup or intrasegmental dipolar interactions of the inter-crosslinked chains, with correlation times $\tau_c > 10^{-4}$ s (they disappear at spinning frequencies less than 2 kHz). These findings suggest that the dynamics of the conformational or segmental motions in crosslinked polymers can only be investigated after the removal of residual static dipolar effects (i.e. with magic angle spinning).

The structure of SL lines, $SL(\omega - \omega_0)$ is explained in terms of a convolution of a residual dipolar broadening function $S(\omega - \omega_0)$ with a narrow Lorentzian line-shape function, $I(\omega - \omega_0)_{MAS}$, originating in the absence of residual dipolar broadening (obtained by magic angle spinning), which is defined by rapid internal motions. The shape of the dipolar broadening function (Gaussian or Lorentzian) does not have a significant influence on the line shape in the wings of the SL lines, because the corresponding line widths obtained by magic angle spinning NMR are much smaller than in the conventional spectrum. The results with Lorentzian shape, $S(\omega, A_L)$, are presented in Fig. 21. The broadening parameter A_L is characteristic of the geometrical restrictions on segmental motions imposed by the crosslinks[3, 113].

The interpretation of the SL line shapes in terms of molecular structure and dynamics is made with the assumptions that a) the mobility of the crosslink

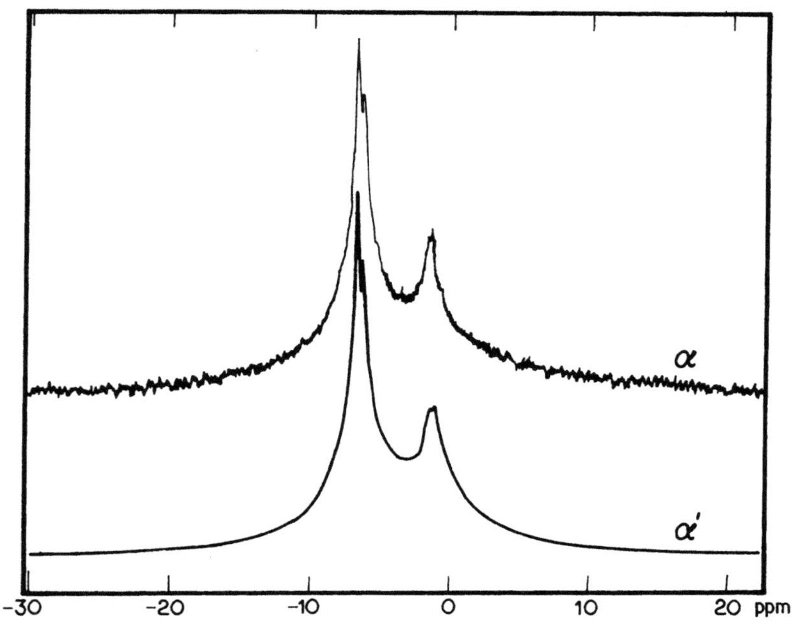

Fig. 21. Standard 100 MHz high resolution NMR spectra of crosslinked polystyrene equilibrium swollen in CCl_4; 65 °C. (α) experimental, (α′) calculated (adapted from Ref. [113])

point is substantially less than that of the chain segments and the inter-crosslink vectors $\mathbf{h}_i$ are defined by their equilibrium average positions (Fig. 22), b) the inter-crosslink chains are so long that the segments in the immediate vicinity of the crosslink points can be neglected, c) the mobilities of all the segments in one inter-crosslink chain are equal and d) the motional behavior of all inter-crosslink chains is the same [111]. Under these conditions, the correlation function for a single inter-crosslink chain, i, is described using a modified model [59]:

$$G_i(\tau) = G_{1i}(\tau)\, G_{2i}(\tau) \tag{33}$$

where G_{1i} corresponds to the spatially restricted rapid segmental motions and is given by:

$$G_{1i} = \Delta M_{2ai} \exp\left(-|t|/\tau_a\right) + \Delta M_{2bi} \tag{34}$$

and $\Delta M_{2i} = \Delta M_{2ai} + \Delta M_{2bi}$ is the rigid-lattice second moment in terms of frequency, $G_{2i}(\tau)$ corresponds to all other (much slower) motional processes and has a single exponential form (Eq. (14)).

The residual part, ΔM_{2bi}, arises from the non-zero average residual projection of the internuclear vectors $\mathbf{r}_{jk}$ in the direction of the inter-crosslink vector $\mathbf{h}_{ij}$ induced by the motional restrictions in a three-dimensional network:

$$\Delta M_{2bi} = |3\cos^2\theta_i - 1|\, \frac{3}{8}\, \gamma^2 h^2 \left[\sum_k \overline{\frac{(3\cos^2\beta_{jk} - 1)^2}{\mathbf{r}_{jk}^6}}\right]^{1/2} \tag{35}$$

ΔM_{2bi} is a function of a) the orientation of the inter-crosslink vector $\mathbf{h}_i$ with the respect to the magnetic field $\mathbf{B}_0$ and b) the time-averaged value of the sum over k, which is the actual measure of the motional restrictions induced by crosslinking. In lightly crosslinked networks, presented by the freely-jointed model of the polymer chain [108], the residual part also can be described by the number of statistical segments in the chain section between crosslinks (Z) (Eq. (24)):

$$\Delta M_{2bi} = \Delta M_{2i} \left(\frac{3}{5Z}\right)^2 \tag{36}$$

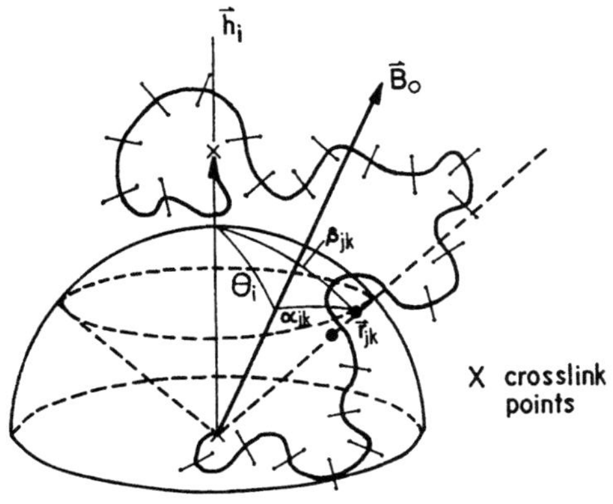

Fig. 22. Schematic diagram of coordinate system related to one inter-crosslink chain (adapted from Ref. [3])

However, the Lorentzian form of the dipolar broadening function, which has the advantage of mathematical simplicity, is not suitable for an interpretation in terms of second moments; it is replaced with a Gaussian dipolar function $S(\omega, A_G)$, where the parameters A_G correspond to the appropriate fractions of the square root of the intra-group rigid lattice second moments. With appropriate values for A_G, calculated and experimental line shapes $I(\omega)$ are found to be in a good agreement for crosslinked poly(ethylene oxide) swollen in chloroform [111].

The dipolar broadening parameter A_G, independent of the type of solvent used, is related to the residual second moment (ΔM_{2b}):

$$\Delta M_{2b} = \frac{\Delta M_2}{4} \overline{(\cos^2 \beta_{jk} - 1)^2} = \frac{4}{5} A_G^2 \tag{37}$$

which describes the crosslinking density in a qualitative and quantitative way. A_G differs for various types of crosslinking agents at same nominal degree of crosslinking (difference in network structure), and it is linearly dependent on crosslink density, i.e. on the reciprocal mean number of carbon bonds between crosslinks points (n). This behaviour is found in swollen polystyrene gels crosslinked with ethylene dimethacrylate (EDM) and divinylbenzene (DVB), up to relatively high degrees of crosslinking (10 and 8 mole%, respectively) [114]. Experimental data for networks crosslinked to these extents could be correlated with a model using the bond rotational potentials rather than the freely jointed model. Since the motional process responsible for the reduction of ΔM_2 has the character of rapid segmental motions (not much affected by the crosslinks), the angular term in Eq. (37), which is a measure of the spatial motional restrictions induced by the crosslink points, can be replaced by the statistical average over all possible conformations of the intercrosslink chain [115]. A statistical mechanical treatment of PS gels yields the relation for the statistical average $(3 \cos^2 \beta - 1) = 10.17/n$, if n has sufficiently large values. The theoretical and experimental data (Fig. 23) are of the same order of magnitude (differences could arise from the above mentioned simplifications and approximations) and indicate the similarity in conformation of inter-crosslink chains in the equilibrium-swollen state and free polymer chains of equal length in dilute solution.

Theoretical relationships between proton and deuterium line widths and shapes in swollen polymer networks and molecular parameters of the system (degree of crosslinking, irregularity of network structure) were also made using freely jointed models with regular and irregular network structures (Fig. 24). Regular network structures (Fig. 24a) are characterized by a constant number of statistical segments Z in a chain connecting two neighboring knots (chain lengths), and there are no pendant chains or chain ends outside the network. If the assumption is made that on the NMR time scale all chain ends are stationary, there can only be small scale anisotropic motions around the axis determined by the vector connecting the chain ends, which do not average dipole-dipole interactions to zero [116]. Thus, for an isolated pair of spins in the weak interaction limit, the anisotropic rotation causes narrowing of the separation between proton doublets by a factor 5/3z and, if the individual line widths are of the same order of magnitude as that of the doublet splitting (ΔH), a bell shaped curve with a width of the order of ΔH is observed. In the case of irregular network structures (Fig. 24b) where some chains have only one end included in the

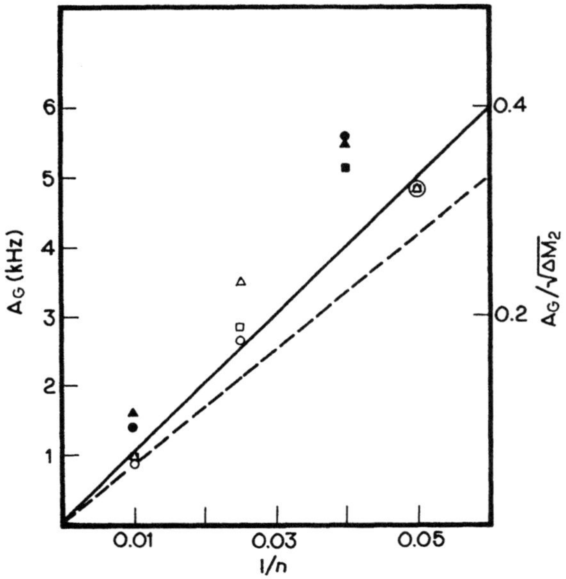

Fig. 23. Plot of the dipolar broadening parameter A_G (left scale) and its relation to the square root of rigid lattice second moment, $\Delta M_2 = 15$ kHz (right scale), for polystyrene networks crosslinked with DVB (solid symbols) and EDM (open symbols) swollen to equilibrium, vs 1/n, the reciprocal "nominal" number of C—C bonds between crosslinks points. Solvents: CCl_4 (□), $CDCl_3$ (△), C_6D_6 (○). (———) linear fit for gels crosslinked with EDM; (— — —) plot of the theoretical value (based on Eq. 37) for statistical average 10/n (adapted from Ref. [114])

network, the number of statistical segments Z is expressed by Eq. (24); if n = 0 the other chain ends are free. Using the appropriate equations for line shape functions, the experimental spectrum of crosslinked PS (with 2 mole% of DVB) is found to be in a good agreement with the model calculated assuming irregular network structure with particular structural parameters. It seems that irregularity or deviation from ideal network structure results in a peculiar lineshape with broad wings and relatively narrow central peak [117].

The same line shapes are observed in deuterium NMR of uniaxially elongated crosslinked rubbers [118]. On the basis of quadrupole effects (taking into account that

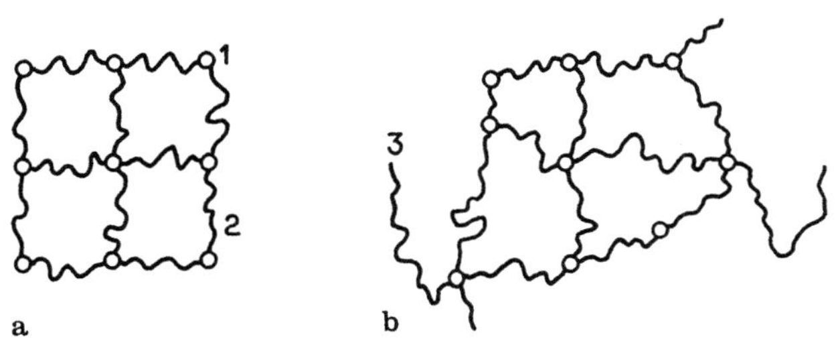

Fig. 24a and b. Schematic simplified representation of the basic difference between a regular **a** and irregular **b** network: knots (1), chains connecting nearest-neighbor knots (2), pendant chains (3)

quadrupole splitting of the CD bond is not influenced significantly by neighboring deuterons), good agreement between experimental data and theoretical considerations is obtained assuming irregularities in the network structure [117].

The effect of the solid-like response of nuclear spins linked to strongly entangled polymer chains in melts and networks on non-exponential free induction decays is explained by the submolecule model, i.e. the chain segments connecting two neighboring entanglements or crosslinks [119, 120, 121]. According to this model, any entangled polymer chain is described as a freely jointed chain and it is divided into "NMR submolecules" of equal contour length characterized by its end-to-end vector **r**. Any submolecule observed on a NMR time scale is supposed to have fixed ends. With the assumption of some residual energy of the spin-spin interactions, the following expression for relaxation function G(t) is established:

$$G(t) = \exp(-t/T_2)\left(\frac{1 + 3\theta^2 + [(1 + 3\theta^2)^2 + 4\theta^6]^{1/2}}{2[(1 + 3\theta^2)^2 + 4\theta^6]}\right)^{1/2} \tag{38}$$

where $\theta = t(M_2)^{1/2}/3$, and second moment M_2 is given by:

$$M_2 = \frac{0.27\gamma^4 h^2}{N_e^2 b^6} \tag{39}$$

where N_e is number of links of a NMR submolecule and b is the distance between two nuclei. Hence, the FID is expressed as a contour length function, with two time ranges; for large time values ($\theta \gg 1$), G(t) behaves like $\theta^{-3/2}$ and the corresponding Fourier transforms I(ω) exhibit SL lines. The experimental decay obtained from spin-echo experiments on entangled PDMS chains in highly concentrated chloroform solution are simulated with Eq. (38), modified for the three spin system. The theoretical curves fit the experimental data and the calculated number of links per submolecule and corresponding apparent average chain molecular weight between coupling junctions are in good agreement with the results obtained from viscoelastic measurements.

Unlike the gels of styrene-divinylbenzene copolymers, crosslinked styrene-ethylene dimethacrylate (St-EDMA) copolymer and linear styrene-methyl methacrylate copolymers exhibit Lorentzian line shapes in the high resolution ^1H NMR. Although there is no evidence for the SL line shape, the integrated intensity of the high resolution signals is smaller than the stoichiometric mass of the copolymer. For the copolymers prepared with 5 and 60 wt% of EDMA, only 70 and 30% of the protons are detectable, respectively, indicating that from the point of view of molecular mobility, St-EDMA in CCl_4 behaves as a two-phase system: one phase is formed by the compact structures inside the microgel-like particles, and the other phase is formed by the "free" copolymer, corresponding probably to St-EDMA segments on the surface of the microgel particles. A comparison with other techniques shows that the composition of the gel fraction is the same, thus allowing the quantitative analysis of the absorption areas. It is found that intramolecular cyclization strongly predominates in crosslinking polymerization yielding, especially at higher EDMA content, compact structures with the character of microgel-like particles. The mechanism of network formation considering changes in the copolymer structure (from NMR data) is explained using a simple statistical polymerization model [122,123].

Changes in intensities and linewidths of ^1H NMR spectra during the hardening of epoxy resins with piperidine, due to the disappearance of the reacting functional groups and the reduced segmental mobility, are used in the analysis of the reaction mechanism [124]. The kinetics of the curing at low temperature of siloxane rubbers was investigated on the basis of differences in the chemical shifts of the protons for the CH_2 groups in initial $(C_2H_5O)_4Si$ and C_2H_5OH formed during the crosslinking process. The activation energy of this reaction is reported as 71 kJ/mole [125]. The kinetics of the crosslinking polymerization of acrylamide (AA) and *N*,*N*'-methylene-bis-acrylamide (BA) is studied at room temperature "in situ" by measuring the changes in the absorption areas of the comonomer signals. The calculated conversion of monomers indicates that the reaction nearly stops after gelation without reaching 100%. At any time BA has greater reactivity than AA. However, the increase in the microviscosity during the reaction results in very broad lines for some network protons (which are usually lost in the noise); even at 90 °C only about 30% of the crosslinked polymers are observed [126].

However, the characterization and concentrations of the functional groups in highly crosslined polymers obtained from plasma polymerized ethylene, acetylene, butadiene gases and benzene vapor are determined from ^1H NMR spectra of CCl_4 solutions (approximately 4 wt. %). All networks show an aromatic proton absorption; aliphatic and vinyl proton resonances are also observed, although the vinyl signal is broad and weak. Quantitative studies are based on the integrated areas, assuming that each aliphatic and —C=C— group is associated with an average of 2.0 hydrogen atoms, respectively. Polymers derived from ethylene and ethylene-acetylene mixtures are found to be nearly identical and contain substantially greater aliphatic character than the ones derived from butadiene and benzene. The increasing aromatic character is consistent with the probability of the phenyl group forming from each monomer, and the results are in agreement with the corresponding IR measurements. It can be concluded that aromatization reactions are suppressed during the polymer formation while branching and crosslinking, yielding a crosslink density of one crosslink per six to ten backbone carbon atoms, are enhanced [127].

A number of ion-exchange resin — solvent systems were investigated by high resolution proton magnetic resonance, mainly from the point of view of the physical chemistry of exchangers [128-131]. The investigated resins were usually based on sulfonated crosslinked polystyrene-divinylbenzene systems and sometimes on resins prepared by the acid condensation of *p*-phenolsulfonic acid with formaldehyde. The aqueous suspensions of these resins exhibit two peaks corresponding to the protons of the water molecules inside and outside the resin, respectively, with the chemical shift difference being proportional to the molality of the resin. The dependenance of the hydration number and the line widths on the degree of crosslinking (% of DVB) were also observed. The difference in chemical shifts between the two types of protons in the water molecules is found to be dependent on the particle size and, consequently, on the degree of crosslinking [132]. With decreasing particle size the spacing decreases, and the two peaks gradually merge until only a single peak remains. The chemical shift of the single peak moves slightly towards lower field with the diminution of the particle size, reaching ultimately a limiting value, which depends on the %

of DVB. With a degree of crosslinking of 11 mole% of DVB, separate signals of the internal and external water are observed at room temperature for particles with a diameter larger than 10μ. In order to obtain resolution of the liquid signals in the corresponding regions of macroporous resins (with higher degree of crosslinking and particle sizes less than 1μ) it is necessary to use more slowly diffusing swelling liquids (methanol, ethylene glycol), as well as lowering the temperature of measurements. From the intensities of these signals the relative volume of pores in the swollen resin was determined, while the chemical shift data are used for the determination of the internal molality and crosslink density [133]. The influence of the solvent on the line widths of polystyrene with bound peptides was also investigated [134].

4.2 High Resolution ^{13}C NMR

Although crosslinked polymers and polymer gels are not soluble, the spectra of swollen, low crosslink density networks exhibit reasonably narrow C-13 NMR line widths, sufficiently resolved to reveal details of microstructure [135]. Thus, recording the spectra under scalar low power decoupling yields characterization information and some dynamic measurements, concerning T_1, T_2 (line widths) and nuclear Overhauser enhancement (NOE) for lightly crosslinked polymers.

4.2.1 Characterization of Polymer Networks

The linewidths in the ^{13}C NMR spectra of swollen chloromethylated polystyrene and poly[styrene-co-(chloromethyl)styrene] [P(S-co-ClS)] crosslinked with divinylbenzene (DVB) are sensitive to the degree of crosslinking. Although all the networks have the same nominal degree of crosslinking (1–2 wt.% of DVB), the backbone methine carbon line widths are the most sensitive measure of the effective degree of crosslinking. Its line widths are found to be greater in chloromethylated polystyrenes than those of P(S-co-ClS), probably owing to the methylene crosslinking introduced during the chloromethylation process [136].

In the series of PS-DVB gels with crosslink densities from 0.007 to 0.112 (as determined by swelling measurements) it was found that the line widths of aromatic carbons measured in $CDCl_3$ increase monotonically with crosslink density. From the log-log plot of these data an empirical exponential relationship between line width ($\Delta\nu$) and crosslink density (χ) can be made [137]:

$$\Delta\nu = 6 \cdot 10^4 \chi^{1.6} \tag{40}$$

The theoretical treatment of line widths and intramolecular dipole-dipole interactions leads to an exponential dependence of $\Delta\nu$ on χ, with the exponent 0.25 (fourth root) if the length of the crosslink is less than the average chain length between crosslinks [138]. In poly(dimethyl siloxane) gels the predicted exponent is found to be 0.75, but experimental results give slightly smaller values (about 0.67), probably due to the effects of chain entanglements in these particular networks [119].

However, it was noticed that in the ^{13}C spectra of swollen gels not all of the carbons in the polymers contribute to the high resolution spectra. Approximately 55 and 35% of the polymer chain contribute to the spectra in the PVP gels with 0.1 and 1 mole% of crosslinking, respectively [139]. A decrease in signal area of the crosslinked PS aliphatic resonances relative to the aromatic area, as well as to the internal standard is also observed. Since this change is not a consequence of increases in T_1 of the aliphatic carbon atoms, it was concluded that the crosslinking limited the motion of some of the backbone carbon atoms so that they were dipolar-coupled and the resulting broad signals were not distinguishable from the spectral base line [140].

The effect of dipolar coupling in highly crosslinked networks on the loss of peak areas was investigated for a series of poly[(chloromethyl)styrene]s crosslinked by divinylbenzene (DVB) and intrapolymer Friedel-Crafts alkylation which takes place when the network is fully swollen. The percent of crosslinking is calculated from the chlorine contents of the polymers using two different models: each Friedel-Crafts crosslinked aromatic ring is bound to only one methylene crosslink carbon (FC-1), and the average Friedel-Crafts crosslinked aromatic ring is bound to two methylene crosslink carbons (FC-2). From the comparison of the experimental aliphatic peak areas with peak areas calculated from models of crosslinking, one can determine which carbon atoms are not detected in the spectra of highly crosslinked polymers. If A and B are uncrosslinked and crosslinked repeating units, respectively, the extent of dipolar coupling is assumed to decrease in the following order: a) crosslink repeat units B, 2) uncrosslinked repeat units A in BAB triads, 3) centered in BAA triads and 4) A centered in AAA triads. Polymers crosslinked with different amounts of DVB fit either the model FC-1 or FC-2. Since these models involve exactly 1 or an average of 2 methylene crosslink carbons per FC-crosslinked rings, and the experimental integrations have possible errors of 10%, more specific conclusions concerning the crosslinks distribution can not be drawn from the NMR spectra alone [141].

On the contrary, integrations of the ^{13}C NMR spectra of linear and 1% crosslinked PS accumulated without nuclear Overhauser enhancement (NOE) give ratios of intensities that are consistent with the chemical structure [134]. Quantitative analysis of swollen chlormethylated crosslinked PS was based on the areas under the $-CH_2Cl$ and methine/methylene peaks (Mannat et al., 1980). The NOE effects on the areas under the $-CH_2Cl$ signal in this analysis are corrected using corresponding factors determined from titration analysis. Observation of the differences in chemical shifts between the $-CH_2Cl$ group substituted at *o*-, *m*- and *p*-positions in the phenyl ring indicates that the chlormethylation yields almost exclusively *para*-isomer.

The extents of cure in polymer networks based on tetrafunctional acrylate and methacrylate monomers are studied using the changes in linewidths of incorporated tracer molecules (actual monomer or a non-reacting molecule mixed in at low concentrations prior to the commencement of cure). The ratio of the peak width of the tracer in the crosslinked sample and in the monomer is used as a quantitative measure of its molecular mobility. Certain tracers in highly cured networks show overlapping absorptions with different line widths, indicating structural inhomogeneity of the matrix, i.e. the existence of two distinct domains with different degrees of molecular mobility [142].

In order to investigate the structure of crosslinked polyethylene by ^{13}C NMR the model compound 1,1,2,2-tetra(tridecyl)ethane, i.e. two molecules of n-$C_{27}H_{56}$ linked at the C-14 atoms, was prepared. The tertiary ^{13}C-14 atoms give rise to a resonance signal at 39.5 ppm. Spectra of long chain n-paraffins eicosane ($C_{20}H_{42}$) and hexacosane ($C_{26}H_{54}$) exposed to 5 MGy γ-radiation above their respective melting points exhibit analogous absorptions due to the tertiary C-atoms, as well as a number of additional peaks. This indicates the presence of H-type crosslinks as a major chemical change induced by the radiation [143].

Similar effects are observed in γ-irradiated n-$C_{44}H_{90}$ (530 kGy) in the molten state. Three new structures are identified as a) one-bond crosslinks (H-structure), b) trans-vinylene groups and c) long branches (T- or Y-structure) [144]. However, highly crystalline polyethylene γ-irradiated in the solid state at low doses (up to 40 kGy) yields predominantly the branched Y-structure. A failure to detect the crosslinked H-structure could arise from a) insufficient abundance of crosslinks to give a detectable signal and b) insufficient mobility of crosslinked chains in the polyethylene gel which results in very broad resonance lines, not observable during normal data acquisition in the solution ^{13}C NMR experiment [145].

A large number of alkyd paint binders (oil modified polyesters formed by the reaction of a polyol, a polybasic aromatic acid and a monobasic fatty acid) and drying oils (unsaturated triglycerides) were investigated in solution (uncured materials) and in the swollen state (crosslinked materials), and the assignments of ^{13}C resonances were given. The general feature of the spectra of crosslinked materials is the loss in intensity for certain absorptions indirectly related to the crosslinks (sometimes present as shoulders), line broadening and poorer resolution due to the restriction of the mobility. Although the well defined signals arise only from the flexible part of the network, and the absorptions connected to the crosslink sites can not in principle be detected, data provide some understanding of the curing mechanisms [146–148].

The structure of highly crosslinked unsaturated polyester can be indirectly analyzed by high resolution ^{13}C NMR spectroscopy, after the hydrolytic degradation of polymer network into a copolymer of styrene and fumaric acid. The range of chemical shifts arising due to the sensitivity of the quaternary phenyl carbon in the styrene repeating unit to the tacticity is about 2.5 ppm. Absorptions arising from configurational differences (comonomer distribution) cover the range of about 8 ppm, so it is possible to assign styrene units along the copolymer backbone. Since the quaternary carbons are chemically very similar, the quantitative analysis of areas under the corresponding absorptions is possible. It is found that poly(carboxylic acids) formed by hydrolytic degradation of the polyester based on maleic anhydride, terephthalic acid and propylene glycol cured with styrene consist of the crosslinks of various styrene lengths (monads, diads and higher n-ads) [149].

The same method is used in the determination of the dependence of the average sequence length on the cure temperature for polyester based on fumaric acid, adipic acid and propylene glycol cured with styrene. The average styrene sequence lengths are found to be in agreement with the stoichiometric values in the resin before curing, except for curing temperatures below 70 °C and above 120 °C where deviations are observed [150]. The styrene content in the hydrolyzates of cured polyester based on phthalic anhydride, maleic anhydride and 1,2-propylene glycol (as determined from ^{1}H NMR) is very close to the initial value in the uncured polyester at fumarate/styrene molar ratio $F/S > 1$; at molar ratio $F/S < 1$ the styrene content increases. The amount of three different triads (FSF, SSS and FSS/SSF), as determined from ^{13}C NMR spectra, is found to be nearly equal at the F/S values near 1.0. A small amount of the SSS triads is observed at $F/S > 1$, while the average number of styrene units in $F(S_n)$ F sequences (approximately 4.5) suggests the existence of long styrene bridges in crosslinked polyesters with higher initial styrene content ($F/S < 1$) [151].

High resolution ^{13}C NMR is also used in the determination of the composition of the dispersed phase in cured rubber modified epoxies in order to analyze the chemical structure of the mobile segments [152]. In this case quantitative analysis is possible because the areas under each peak are approximately equal to the number of carbons contributing to the peak, and the intensities of the broad lines from the rigid phase are very low, almost indistinguishable from the baseline noise. The structure of crosslinked networks based on poly(3,4-pyrrolidinediethylene), synthesized by different methods, was determined from gels swollen in water and chloroform [153].

Proteins crosslinked by formaldehyde are important in photography, the leather industry and in bio-medical sciences. Due to the complex structure of the gelatin molecules (consisting of approximately 20 different kinds of amino acids) and the very low crosslink density, it is not possible to detect crosslink resonances under normal conditions. In order to overcome this problem a ^{13}C enriched formaldehyde is used. By comparison with the chemical shifts of model crosslink compounds it is concluded that the predominant crosslink is formed between the lysine and arginine components in gelatin. A possible mechanism for the reaction between these two amino acid components and the formaldehyde has been proposed [154].

The progress of the curing reaction of diglycidyl ether of bisphenol A (DGEBA) with *m*-phenylene diamine, and assignment of new structures (secondary and tertiary amines) was followed using the ^{13}C NMR spectra at 100 °C. The reaction kinetics are followed from the change in intensity of specific carbon signals [155]. From the results obtained from three different techniques, ^{13}C NMR, ^{15}N NMR and high performance liquid chromatography (HPLC), it was concluded that the cure of the system proceeds mainly by chain extension, whereas crosslinking occurs in the reaction of hydroxyl groups with epoxides and results in the formation of ether linkages [156]. An analogous application of those three techniques is found in the studies of polyaddition of diglycidyl-1-methyl benzene, where the role of impurities on reaction kinetics and on the structure of the final polymer networks is investigated [157]. The studies of the reaction between the model diepoxy aromatic compounds and aniline indicate that i) the network build-up is dependent on the reactivity of the functional groups and their position on the aromatic ring, and ii) that the curing reaction is an autocatalytic process characterized by $E_a \sim 62.7$ kJ/mol and a non-catalytic path characterized by a second order reaction [158].

4.2.2 Relaxation in Polymers

The relaxation of the ^{13}C nucleus is dominated by $^{13}C—^{1}H$ dipolar interactions. For slow rotational reorientation ($\omega_c\tau_c \gg 1$) assuming a single correlation time, τ_c, the following equations for ^{13}C spin-lattice and spin-spin relaxation times are valid:

$$\frac{1}{nT_1} = \frac{\gamma_H^2\gamma_C^2h^2}{10r^6}\left[\frac{(\omega_H - \omega_C)^{-2} + 2\omega_C^{-2} + 6(\omega_H + \omega_C)^{-2}}{\tau_c}\right] \tag{41}$$

$$\frac{1}{T_2} = \frac{\gamma_H^2\gamma_C^2h^2}{5r^6}\tau_c \tag{42}$$

In the case of extreme narrowing, $(\omega_C + \omega_H\tau_c \ll 1)$:

$$\frac{1}{T_1} = \frac{1}{T_2} = \frac{\gamma_H^2\gamma_C^2h^2}{\tau_c r^6} \tag{43}$$

where γ_C and γ_H are the ^{13}C and ^1H gyromagnetic ratios, ω_C and ω_H are the corresponding resonance frequencies, and n is the number of protons attached to the carbon atom. The T_1 minimum occurs, analogously to proton T_1, when $\omega_C\tau_c \simeq 1$. However, ^{13}C NMR spectra obtained under the conditions of proton decoupling give rise to a nuclear Overhauser enhancement (NOE) of ^{13}C resonances. The NOE for systems relaxing only by the dipolar mechanism has a value 2.988 in the extreme narrowing region ($\tau_c > 10^{-10}$ s). The NOE decreases in the vicinity of the T_1 minimum to 1.153, and is τ_c independent for slow motions.

Various types of covalently crosslinked gels, including poly(*N*-vinyl-2-pyrrolidone) (PVP), poly(hydroxyethyl methacrylate) (PHEMA) and copolymer of NVP and methyl methacrylate (MMA) are swollen by water [139]. In the gels of PVP with 90, 75 and 53% water content, a significant increase in line widths with increasing crosslinking is observed. Line broadenings are explained in terms of restriction of the chain mobility due to the presence of chemical crosslinks and entanglements of the polymer chain, particularly in the vicinity of crosslinks. Although the network formation has an effect on the T_2 relaxation times, the degree of crosslinking does not

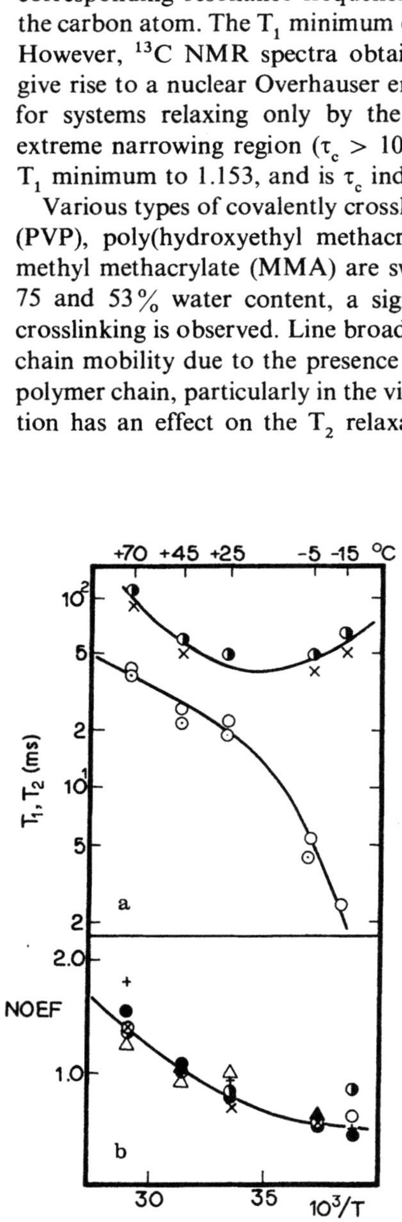

Fig. 25a and b. Temperature dependence of T_1, T_2 **a** and NOEF **b** for polystyrene in CCl_4. **a** aromatic $C_{2,3}$ T_1 for linear (◑) and crosslinked PS (×); T_2 for linear PS aromatic $C_{2,3}$ (○) and methine CH (⊙). **b** NOEF: linear PS aromatic C_1 (○), aromatic $C_{2,3}$ (◑) and methine CH (●); crosslinked PS aromatic C_1 (×), aromatic $C_{2,3}$ (+) and methine CH (△) (adapted from Ref. [113])

affect the T_1 and NOE values. An increase of line widths with the degree of crosslinking is also observed in swollen poly(sucrose methacrylate) gels [159].

The temperature dependence of the relaxation parameters (T_1, T_2, NOE) for swollen polystyrene gels, crosslinked with 2 mole% of divinylbenzene indicates little differences in the dynamics of the rapid motions in linear and crosslinked polymers (Fig. 25) [113]. Both linear and crosslinked PS exhibit the same T_1 minima (higher than the theoretical value predicted for isotropic reorientation). Furthermore, the T_1 and T_2 data suggest the same distribution (or anisotropy) of correlation times, which is not very broad. Although another study of crosslinked polystyrene gels (crosslinked with 0.5, 1, 2, 4, 6 and 10% divinylbenzene) also reports little change in the T_1's of both backbone and aromatic ^{13}C nuclei, an increase in the line widths (T_2) and a significant decrease in the NOE ratios with increasing crosslinking was found [139]. Formation of crosslinks introduces low-frequency motions in the polymer system that affect the T_2 and NOE processes but have little effect on T_1. Furthermore, increasing the degree of crosslinking (up to 12% divinylbenzene) causes restriction of the molecular dynamics of the counter-ion (triethylbenzylammonium) in the polystyrene ion exchange resin phase [160].

The influence of small amounts of crosslinks was examined in swollen syndiotactic-rich poly(methyl methacrylate) (PMMA) lightly crosslinked with triethylene glycol dimethacrylate (TEDGM) and 2,2-bis[4-(2-methacryloyloxy-ethoxy)phenyl]propane (BMEPP). There was no significant difference between the line widths of linear PMMA and crosslinked PMMA with 0.01 mole% of crosslinking agent, while all the line widths increase in the samples with 0.1 to 1 mole% of crosslinking agents [161].

It should be noted that the activation energies for motional processes in the same crosslinked polymer gel calculated from line widths in ^{13}C NMR spectra are higher than those calculated from ^1H NMR spectra under the magic angle conditions (residual line width). These findings indicate that ^{13}C line widths (which are of the order of 10 Hz) are probably more affected by sample inhomogeneity and by relatively small residual chemical shift anisotropies [162].

The investigation of ^{13}C NMR relaxation times over a wide temperature range in the PS-DVB network (2 mole% of DVB) swollen in CCl_4 (15% by weight) show that both linear and crosslinked PS exhibit a pronounced minimum at 285 K (with the minimum value of T_1 40 ms), although the T_1 values for PS-DVB are in all cases slightly less than those of linear PS [113].

Segmental motion of side chain substituents attached to slightly crosslinked polystyrenes results in sharply defined ^{13}C NMR signals in the swollen networks, which enable identification and relaxation studies of the functional groups in these gels. Chemical shifts of the polymer-bound trityl alcohol, monotrithyl ethers of the symmetrical diols $HO(CH_2)_nOH$ (n = 2, 4, 6, 7, 9 and 10) and some related intermediates in the solid phase synthesis of insect pheromones were assigned [163]. Relatively narrow NMR signals provide the investigation of slightly crosslinked polystyrenes (1% DVB) bearing methylenetri-n-butyl-phosphonium chloride or mesylate groups in terms of the pendant group mobility [164], as well as the study of the solvation of quaternary phosphonium and ammonium ions in different solvents [165].

Deuterium-decoupled deuterated amino acids bound to crosslinked polystyrene with sulfonic acid groups show line widths 3 to 4 times narrower than those from proton-decoupled, protonated amino acids, suggesting an anisotropic model for the resin-bound amino acids which yields a nonzero time-averaged dipolar interaction [166].

4.3 High Resolution NMR of Other Nuclei

Among the nuclei which form the main polymer chain, besides carbon, silicon is the next most significant; the natural abundance of the ^{29}Si isotope (4.7%) allows detection of the silicon resonances. The molecular motions in poly(dimethyl siloxane) (PDMS) elastomeric networks at room temperature are sufficiently rapid to narrow the rigid lattice line widths and result in a high resolution spectrum. However, the main problem in the silicon polymers and networks in the fluid state is the long T_1 relaxation times. The T_1 for the PDMS tetra- and tri-functional silicons are approximately 90 s, and in the non-fully relaxed spectra (obtained with less than 8 min delay time between pulses) the intensities of these resonances are significantly diminished. Doping of the networks with a paramagnetic molecule, such as chromium acetylacetonate, results in a shortening of the long T_1 relaxation times from about 90 to approximately 40 s, which is still inconvenient for rapid data acquisition. The difficulties connected with the accurate measurements of dilute network junctions in the fluid state due to the long ^{29}Si T_1 relaxation times are not present in the solid state high resolution ^{29}Si NMR, which is found to be much more useful for characterizing silicon elastomeric networks at low temperatures [167].

A series of linear and crosslinked polymers, synthesized from poly(difluorophosphazene) with alkyl lithium reagents, have been characterized by both high resolution ^1H and ^{31}P NMR spectroscopy in the swollen state. Resonances arising from ^{31}P are assigned to the phosphorus atoms bonded with alkyl groups (R) and trifluoroethoxy groups (F) in the three possible repeating configurations: $-N{=}PR_2^-$, $-N{=}PRF$- and $-N{=}PF_2$-. The ratios R:F are determined from proton and phosphorus NMR, and the percentages of alkyl groups are found to agree within 5% by these two techniques. A mechanism for the reactions of the crosslinking is also proposed [168].

5 High Resolution Solid State NMR

The broad line spectra of nuclei with spin $I = 1/2$ in the solid state are mainly a consequence of the dominant contribution of the dipolar Hamiltonian $\mathbf{H}_D$ (Eq. (4)), which gives rise to a local field $\mathbf{B}_{loc}$. Its magnitude varies as a function of the angle θ_{ij} between the internuclear vector $\mathbf{r}_{ij}$ and the applied magnetic field $\mathbf{B}_0$. Depending on the nature of spin system, two general types of interactions can be distinguished:

1. Homonuclear dipolar interactions described by the term $\mathbf{H}_D$ are present in the spin system of abundant nuclei, such as ^1H and ^{19}F:

$$\mathbf{H}_D = \frac{\gamma^2 h^2}{4} \sum_{i<j} \frac{(1 - 3\cos^2\theta_{ij})}{\mathbf{r}_{ij}^3} (3\mathbf{I}_{iz}\mathbf{I}_{jz} - \mathbf{I}_i\mathbf{I}_j) \tag{44}$$

2. In the "diluted" spin system such as ^{13}C and ^{29}Si, homonuclear dipolar interactions are negligible because of the low natural abundance of nuclei under observation (S); in this case term $\mathbf{H}_D$ describes dipolar interactions between unlike spins (I and S), which for ^{13}C and ^{29}Si are most frequently the neighboring protons:

$$\mathbf{H}_D = \gamma_I \gamma_S h^2 \sum_{i,j} \frac{(1 - 3\cos^2\theta_{ij})}{\mathbf{r}_{ij}^3} \mathbf{I}_{iz}\mathbf{S}_{jz} \tag{45}$$

Owing to the differences in both the nature and magnitude of the homo- and hetero-nuclear dipolar interactions, the process of eliminating the dipolar interactions to obtain a high resolution spectrum in the solid state is slightly different for protons and carbons, respectively, and will be reviewed separately in following sections.

One common technique for reducing resonance line widths, (used in both cases), is magic angle spinning (MAS). The dipolar Hamiltonian $\mathbf{H}_D$ can be reduced to zero if the contribution of the term $(1 - 3\cos^2\theta)$ is zero [169,170]. Hence, mechanical spinning of the sample at the angle $\theta = 54°44'$ (the so-called magic angle) causes an analogous effect to the rapid molecular motion in solutions, i.e. narrowing of lines and high resolution isotropic spectrum. Furthermore, rotation of the solid sample about the magic angle averages the chemical shielding tensor $\boldsymbol{\sigma}$ (Eq. (30)) to its isotropic value, thus further narrowing the line.

However, eliminating $\mathbf{H}_D$ by MAS is successful only when the spinning frequency is greater than the interaction to be averaged. MAS high resolution spectra are obtained for particular spin systems of abundant nuclei, such as protons in swollen crosslinked polymers, which have a) relatively small average coupling constant (J_{HH}) and b) a certain degree of molecular motion (τ_c of the order 10^{-4} s or longer). Although line-narrowed high resolution spectra are recorded for these protons, MAS is generally not applicable for removing much larger heteronuclear dipolar interactions of rigid rare spins. Limitations arise in obtaining spinning speeds with frequencies faster than the magnitude of the dipole-dipole couplings. Typically, spinning frequencies in the range of 4 to 5 kHz are not sufficient to suppress ^{13}C—^{1}H dipolar couplings in solids (approximately 50 kHz).

Hence, originally designed to eliminate dipolar interactions, MAS has the main effect on the removal of the chemical shift anisotropy, which is rather pronounced in ^{13}C NMR spectroscopy. Since heteronuclear interactions could be viewed as a perturbation of abundant spins (I) on the energy states of rare spin system (S), a more convenient way of reducing $\mathbf{H}_D$ to zero is high power decoupling of the I nuclei ($\mathbf{I}_{iz} = 0$).

The combined use of several different NMR techniques including a) magic angle spinning, b) homonuclear and heteronuclear decouplings, c) multiple pulse sequences and d) cross-polarization, gives rise to a large number of new possibilities for the investigation of the structural and dynamical properties of materials in the solid state particularly for insoluble crosslinked polymeric networks.

5.1 High Resolution Solid State ^{1}H NMR

The line widths in NMR of solid materials, $\Delta\nu$ (usually of the order 40 to 60 kHz), are governed by the spectral density function term $J_0(0)$, which dominates the equation for the spin-spin relaxation time T_2 (Eq. (10)) because of the very slow internal motion ($\omega_0\tau_c > 1$). Mechanically induced spinning of the sample, with the spinning frequencies ν_r usually in the kHz range, is approximately four or five orders of magnitude smaller than the resonance frequency ν_0. Consequently, spectral density functions $J_i(\omega_0)$, which determine T_1, are not affected by the spinning frequency ν_r (or expressed in the terms of angular frequency $\omega_r = \nu_r 2\pi$). However, a significant line narrowing, i.e. change in T_2 due to the effect of ω_r on $J_0(0)$, can be achieved if the spinning frequency and correlation time of the internal motion satisfy the relation $\omega_r^2\tau_c^2 > 1$. An example of NMR spectra of linear and swollen crosslinked polystyrene, respectively, measured conventionally and with magic angle spinning (MAS) is shown in Fig. 26. The limiting line width is reached at a spinning frequency of about 1.5 kHz.

Since crosslinked polymer networks exhibit anisotropic internal motions, the effect of MAS on the line narrowing is explained assuming that the rigid lattice second moment ΔM_2 consists of two terms, analogous to Eq. (34). Then, the residual line width in the NMR MAS experiment, $\Delta\omega_{MAS}$, can be described by the following relation:

$$\Delta\omega_{MAS} \sim \Delta M_{2a}\tau_{ca} + \left(\frac{\Delta M_{2b}}{\omega_r}\right)^2 \frac{1}{\tau_{cb}} \tag{46}$$

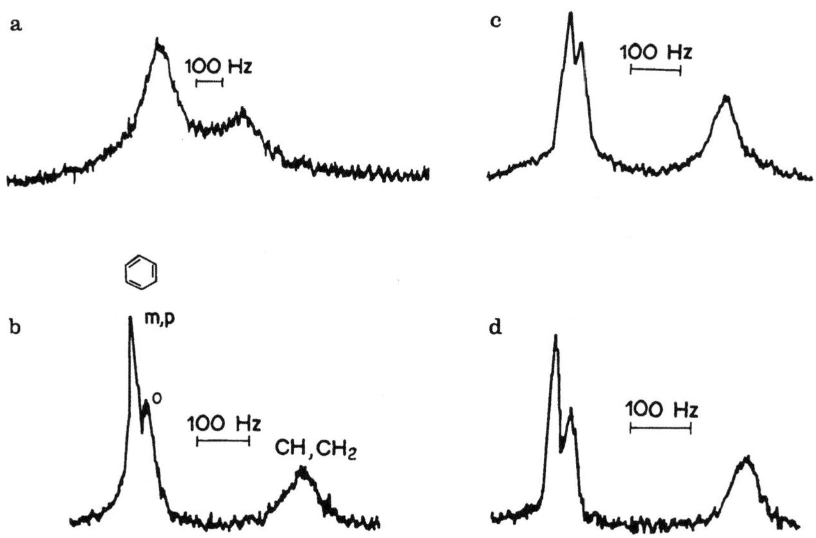

Fig. 26a–d. ^{1}H NMR spectra of polystyrene in CCl_4 (20 °C, 60 MHz): **a, b** crosslinked through *ortho* (*o*), *meta* (*m*) and *para* (*p*) positions, equilibrium swollen, **c, d** linear; **a, c** standard high resolution **b, d** MAS (reprinted from Ref. [3]) with permission)

where ΔM_{2a} is part of the rigid lattice second moment (in terms of angular frequency ω) removed by random microscopic motion characterized by the correlation time τ_{ca}, and ΔM_{2b} is the part removed by macroscopic motion (MAS); time τ_{cb}, associated with any motion capable of removing ΔM_{2b} microscopically, is too long to be effective ($\tau_{cb} \gg \tau_{ca}$)[171].

Molecular motions in polymers, particularly in the solid state and in the polymer networks, are still not well understood in terms of the types and extents of motions, as characterized by the activation energies and the distribution of correlation times. Different effects of crosslinking on high- and low-temperature motional processes in various polymer systems are observed from broad line and pulsed NMR (the calculated E_a are listed in Tables 1 and 2). Since the residual line widths in the MAS experiment are proportional to the effective τ_c of rapid internal motions, it is possible to estimate activation energies for motional processes in swollen polymer networks from the linear dependence of the line widths on the reciprocal of temperature, $E_a = (R \Delta \ln \Delta\nu_{MAS})/\Delta(1/T)$. The E_a values for both proton groups in lightly crosslinked polystyrene (PS) (2 mole% of ethyleneglycol dimethacrylate (EDM) and swollen in CCl_4 to equilibrium are approximately the same (14.0 kJ/mole). For highly crosslinked PS (5 mole% od EDM) exhibit for aromatic and aliphatic protons, $E_a = 21.9$ and 31.5 kJ/mole, respectively, indicating corresponding restrictions of the rapid segmental motion imposed by the crosslinking[172]. Similarly, the E_a (calculated from the temperature dependence of line widths ($\Delta\nu_{MAS}$) and Eq. (46)), and line widths of backbone methylene groups are higher than those of α-CH_3 and —OCH_3 side groups at the same degree of crosslinking. All proton groups in linear PMMA have the same E_a, which slightly increases with increasing crosslink density up to 1 mole% of crosslinking agent. This suggests that the restriction of molecular motion in swollen PMMA gels due to the crosslinking is most pronounced in the polymer backbone[112]. Activation energies for motions in the swollen crosslinked poly(ethylene oxide) (PEO) gels depend on the particular polymer-solvent system and are 7.1 and 39.3 kJ/mole in chloroform and CCl_4, respectively. These differences are probably due to the effects of local conformation and viscosity in these solvents[162].

The $\Delta\nu_{MAS}$ for swollen crosslinked PS-EDM increases with increasing degree of crosslinking (from 2 to 10 mole%), probably indicating a broadening of the distribution of correlation times in the direction of slower modes. Line widths are also sensitive to the type of solvent; their higher values in thermodynamically poor solvents (i.e. CCl_4) arise from less mobility as compared to good solvents like chloroform and benzene. However, the temperature dependence of $\Delta\nu_{MAS}$ for the aliphatic and aromatic protons in a swollen network with 2 mole% of EDM does not exhibit a linear dependence on $1/T$ and, therefore, they cannot be interpreted with a single τ_c and activation energy. Since solutions of linear PS of equal solvent concentration show the same temperature dependence of $1/T_2$, it can be concluded that the same motional mechanisms, characterized as rapid segmental motions (short-range conformational motions), are governing both $\Delta\nu_{MAS}$ of crosslinked gels and $\Delta\nu$ of the linear polymer. Their frequencies are almost independent of the degree of crosslinking[114].

By means of both conventional and MAS high resolution NMR, a characterization of macroporous ion-exchange resins were characterized (PS-DVB with 2, 4, 8, 15 and 25 nominal % of DVB) was also made[173,174].

The magic angle spinning technique applied to crosslinked polymer systems yields valuable information about network dynamics, because of the possibility of observing the effect of specific anisotropic molecular motions. In addition, knowledge of $\Delta\nu_{MAS}$ and the corresponding Lorentzian line shape functions $I(\nu - \nu_0)_{MAS}$ enables qualitative and quantitative analysis of line shapes in conventional NMR of molecules undergoing anisotropic motions (Sect. 4.1). However, the general application of this technique to solid polymers is limited by the inability to obtain sufficiently fast spinning speeds to efficiently remove dipolar interactions.

An alternative technique of line narrowing is the application of multiple pulse sequences. Multiple pulse sequences, such as the two pulse Hahn spin echo [83] and the cyclic two pulse Carr-Purcell sequence [175] are commonly used in conventional pulsed NMR for measuring line widths much narrower than the magnetic field inhomogeneity. Application of special pulse sequences to solids results in an artificial time dependent Hamiltonian $\tilde{H}(t')$ which reflects the effect of a sequence of n pulses on the suppression of homonuclear dipolar broadening. Using the concept of an average Hamiltonian, four main characteristics of the multiple pulse sequences used to obtain averaging of internal interactions can be determined: a) the sequence must be cyclic, b) the cycle time t_c must be short compared to the T_2 characterizing the homogeneous broadening due to the averaging interaction, c) groups of sequences must be periodic and d) the response of the spin system must be observed at "windows" which are integral multiples of the t_c [176].

The main purpose of the sequences is to obtain an averaged Hamiltonian $\bar{H}$ in which the dipolar term is very small compared with the chemical shift term. The zeroth-order of the average dipolar Hamiltonian term is given by the following equation:

$$\bar{H}_D^{(0)} = \frac{1}{t_c} \int_0^{t_c} \tilde{H}_D(t')\, dt' \tag{47}$$

where $\tilde{H}_D(t')$ describes the effect of a particular pulse sequence on the dipolar Hamiltonian. For example, two 4-pulse sequences, WHH-4 (WAHUHA) [177] and MW-4 [178, 179] with the pulse structures:

$$\text{WHH-4:} \quad (\tau - P_{-x} - \tau - P_x - \tau - P_y - \tau - P_{-y} - \tau -)_n$$
$$\text{MW-4:} \quad (\tau - P_x - 2\tau - P_x - 2\tau - P_x - 2\tau - P_x - \tau -)_n$$

following a P_{-y} preparation pulse, where P_i are ideally δ pulses with a rotation angle of 90°, have different effects on $\bar{H}_D^{(0)}$ and $\bar{H}_\delta^{(0)}$. Spectra of swollen crosslinked polystyrene gels investigated with these two multiple pulse sequences (as well as with magic angle spinning and conventional NMR) are shown in Fig. 27 [180]. The WHH-4 pulse sequence completely averages the dipolar interactions and the chemical shift scale is reduced by a factor $\sqrt{3}$. A sequence of P_x pulses in MW-4 completely averages out the chemical shift term H_δ ($\delta = 0$) and reduces the line broadening typical for solids (Fig. 27d). The relaxation time T_{2eff} in the MW-4 experiment depends on the correlation time τ_c, wnich is generally different for the various protons. Consequently, only the mean T_{2eff} values are observed, which exhibit a minimum, analogously to the $T_{1\varrho}$ when the effective radio frequency field and τ_c are of the same order

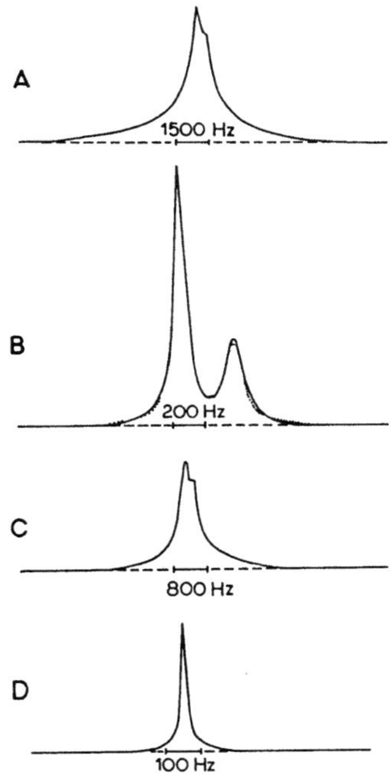

Fig. 27. ^{1}H NMR spectra of crosslinked polystyrene gel swollen in $CDCl_3$: (A) Fourier transform of FID, (B) CW with magic angle spinning [— experimental; ... calculated], (C) WHH-4 pulse sequence, (D) MW-4 pulse sequence (adapted from Ref. [180])

(typically in the millisecond region). With increasing crosslinking, $T_{2\,\mathrm{eff}}$ is shifted to higher temperatures, indicating restrictions in the molecular dynamics [181]. The chemical shift analysis with the MW-4 pulse sequence is not possible, but this multiple pulse experiment gives useful information from the line shapes.

Linear and carbon-filled, sulfur vulcanized natural rubber were investigated using the eight-pulse sequence MREV-8, which has the following structure [182,183]:

$$(\tau - P_x - \tau - P_{-y} - 2\tau - P_y - \tau - P_{-x} - 2\tau - P_x$$
$$- \tau - P_{-y} - 2\tau - P_y - \tau - P_x - \tau \; -)_n -$$

A modified version was used by introducing a phase error in the P_x pulse in order to remove the effects of the chemical shift and magnetic field inhomogeneities [184,185]. The line shape of the multiple pulse spectrum contains contributions from the chemical shift Hamiltonians, relaxation phenomena and instrumental effects. By applying both versions of MREV-8 pulse sequences, it is possible to isolate the contribution due to the chemical shift Hamiltonian. Natural rubber exhibits an exponential decay envelope with the time constant of 17 msec while filled vulcanized samples show composite decays of two Lorentzian curves (analogous to those shown in Fig. 11) with time constants of 1.5 and 13.5 msec for the short and long decay component, respectively. The relative contribution of the long decay of about 40% corresponds to the relative fraction of methyl protons (37.5%); the short component is attributed to the backbone protons. The frequency spectrum of the methyl group motion is less altered by the crosslinking process, because they have substantially more motional freedom than the backbone protons (attached to the carbons which are an integral part of the molecular frame).

A number of different multiple pulse sequences (8-, 24- and 52-pulse sequences) have also been introduced in order to obtain better resolution or line narrowing, i.e. to affect the first- and second-order terms in the average Hamiltonian. Since pulse imperfections are the major source of resolution limitations, these composite pulse sequences are designed with corresponding symmetry properties which allows the canceling of specific rf pulse imperfections.

In general, multiple pulse techniques sufficiently average the dipolar interactions, compress the chemical shift scale, but they do not affect heteronuclear dipolar interactions and the chemical shift anisotropy. A combination of both multiple pulse techniques and magic angle spinning, so-called CRAMPS (Combined Rotational And Multiple Pulse Spectroscopy) is found to yield satisfactory results in the solid state ^{1}H NMR of solids [186]. The limitations of all these techniques, from the analytical point of view, arises from the relatively small chemical shift range (about 10 ppm) as compared with some other frequently studied nuclei. However, high resolution ^{1}H NMR of solids is useful in studies of molecular dynamics.

Although the multiple pulse techniques and/or magic angle spinning are necessary for obtaining high resolution proton NMR spectra of solid glassy polymers, the line width of rubberlike materials is so narrow that the liquid state conditions can be used to study them even in the solid state. The changes in the linewidth ($\Delta\nu$) of the polychloroprene rubber —CH= absorption vs. temperature is followed as a function of curing time. It is found that the $\Delta\nu$ decreases above 120 °C, indicating that network degradation predominates over network formation, and the activation energy for the degradation process is linearly dependent on the crosslink density [187].

5.2 High Resolution Solid State ^{13}C NMR

The carbon-proton dipolar interaction and the chemical shift anisotropies broaden the lines in solid state ^{13}C NMR spectra. The major effect arises from the dipolar coupling of the carbon nuclei with neighboring protons; homonuclear dipolar couplings between two adjacent ^{13}C nuclei are neglegible because of their low natural abundance. The large magnitude of dipolar ^{13}C—^{1}H coupling (up to 40 kHz) results in broad and structureless proton-coupled ^{13}C NMR absorptions.

The heteronuclear dipolar broadening can be removed by the irradiation of protons with a strong radiofrequency (rf) field, at or close to their Larmor frequency. This decoupling is similar to the scalar decoupling used in high resolution ^{13}C NMR in solutions except that the power level is an order of magnitude higher. The applied irradiation coherently averages the dipolar couplings. The intensity of high power decoupling must be larger than the ^{13}C—^{1}H dipolar coupling strength and the natural proton dipolar linewidth. For example, to reduce a 25 kHz linewidth by a factor of 1000 (25 Hz), an rf field of 68 kHz is required (rf field is usually of the order comparable to 10 to 20 T^{-4}). The application of magic angle spinning (which removes the chemical shift anisotropy) and high power decoupling, allows the recording of high resolution ^{13}C NMR spectra in the solid state. An example of the spectra

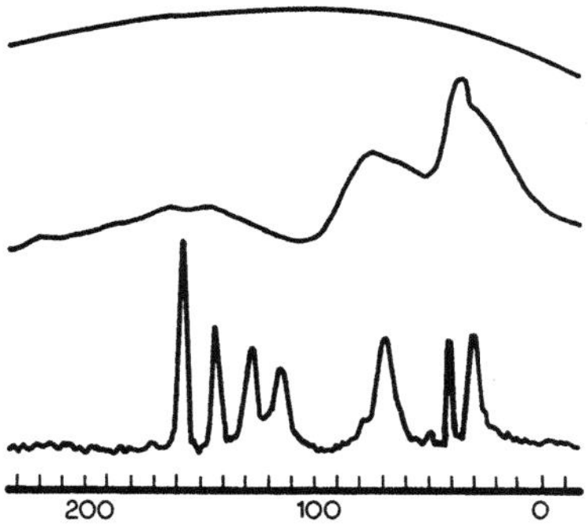

Fig. 28. Three stages of resolution in a ^{13}C NMR spectrum of a cured epoxy. The top spectrum is obtained under conditions appropriate to a liquid-state spectrometer: no dipolar decoupling and no magic angle spinning. Dipolar decoupling at 60 kHz is used for the middle spectrum and to that is added magic angle rotation at 2.2 kHz for the bottom figure (reprinted from Ref. [188] with permission)

obtained using high power dipolar decoupling and the magic angle spinning on "liquid state" NMR conditions for a cured epoxy resin is shown in Fig. 28.

The signal arising from carbon nuclei in the solid state can be generated in two ways: a) directly, by applying a 90° pulse on the ^{13}C nuclei and b) indirectly by building up the carbon magnetization via polarization transfer from the abundant protons (cross-polarization). The signal intensity of the gated or cross-polarization technique, respectively, is mainly determined by the values of carbon T_1 relaxation times. The delay time between the pulse sequences, must be long enough to recover the initial carbon magnetization. In solids the T_1s are long (up to 1000 s) so the gated experiment has insufficient sensitivity. Generally, glassy and crystalline polymers exhibit better spectra if cross-polarization is applied, while the carbons in more mobile polymers, such as elastomers, show good response to both excitations, although the cross-polarization experiment is more difficult in the latter case.

5.2.1 Gated High Power Decoupling

The gated high power decoupling (GHPD) pulse sequence (Fig. 29) involves a 90° carbon pulse and the observation of the FID under the conditions of high power decoupling. This pulse sequence is successfully applied to polymer systems where there is a sufficient degree of mobility and the carbon T_1 relaxation times are not greater than a few seconds. Elastomeric materials in the cured and uncured state have such properties and solid state NMR can be carried out by building up the carbon magnetization directly. An early study of a carbon-black-filled, sulfur-vulcanized, standard tread-stock sample of cis-polyisoprene shows that the line broadening due to

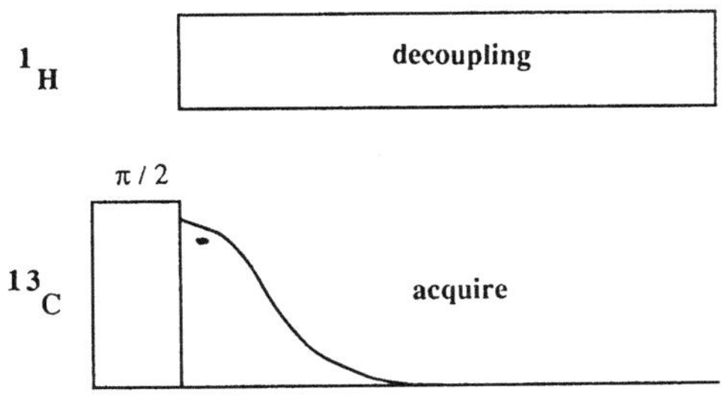

Fig. 29. Gated high power decoupling (GHPD) pulse sequence

anisotropic motion and the residual dipolar interactions could be reduced by a factor of 3–4 by spinning the sample at the magic angle (spinning frequency of 1 kHz is applied with the proton decoupling of $\gamma B_1 = 6.3$ kHz). The narrowing of the ^{13}C resonances permits the resolution of a new absorption (between two methylene carbons). This weak line, not observable in nonvulcanized rubber, was tentatively assigned to the methylene carbons covalently bonded to sulfur and forming a part of the crosslinked network [189]. Owing to a great degree of segmental mobility in elastomers, some studies are possible even without magic angle spinning; enhanced mobility at elevated temperatures is sufficient to remove the chemical shift anisotropy and sufficiently narrow the spectral lines [200].

Elastomeric materials have specific dynamical properties, mainly the anisotropic reorientational motion above the glass transition and, from the point of view of NMR, relatively weak dipolar couplings as compared to the solid glassy polymers. Thus, the line narrowing and the resulting high resolution spectra for this group of polymers can be obtained by applying relatively lower spinning speeds and proton power decoupling than in the standard high resolution solid state NMR method. The effect of different techniques used in high resolution solid state NMR, such as magic angle spinning, multiple pulse and high power decoupling, respectively, on the spectra on an elastomer are shown in Fig. 30 [191]. The line widths of ^{13}C resonances in natural rubber are approximately 400 Hz (Fig. 30a). The broadening by dipolar interaction is one order of magnitude smaller as compared to adamantane, a standard solid state high resolution sample. The high power proton decoupling (Fig. 30b) narrows the spectral lines to 300 Hz, i.e. a contribution of dipolar interaction to the line width in Fig. 30a is only 100 Hz, indicating the presence of fast

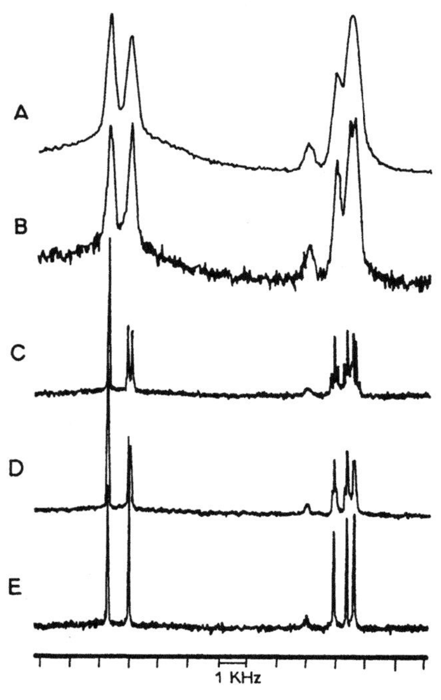

Fig. 30. Spectra of filled natural rubber obtained under various conditions: without MAS and decoupling (A); with high power proton decoupling but no MAS (B); with MAS but no proton decoupling (C); with MAS and multiple pulse MREV-8 decoupling (D); with MAS and high power proton decoupling (E). Resonance at 43 ppm arises from an antioxidant and it is not a subject of the study (adapted from Ref. [191])

(nearly isotropic) motions that average the dipolar interactions and the chemical shift anisotropy. The application of magic angle spinning (MAS) removes the chemical shift anisotropy and narrows the lines one order of magnitude (30 Hz) (Fig. 30c–e). Since all the interactions except for the isotropic chemical shift and the scalar J_{CH} spin-spin coupling (Eqs. (25) and (26)) are averaged to zero, the well-resolved J multiplets are observed (Fig. 30c). The multiple pulse decoupling (with MREV-8 pulse sequence) in the rubber sample, where the dipolar interactions are averaged to a greater extent by the the 2 kHz MAS alone, results only in the scaling of J_{CH} couplings and undesired reduction of the resolution (Fig. 30d). However, it should be noted that the multiple-pulse decoupling, which eliminates only the proton homonuclear coupling (Eq. (39)), is necessary in order to observe the scalar couplings in a typical solid [192]. The high power proton decoupling eliminates both the dipolar broadening (if it is not reduced by MAS) and the heteronuclear dipolar interaction (Eqs. (39) and (40)), resulting in the standard decoupled high resolution solid state ^{13}C NMR (Fig. 30e). The fact that the MAS at the moderate frequency of 2 kHz narrows the lines from 300 to 30 Hz indicates the presence of restricted motions and residual dipolar couplings, caused by the spatial restrictions introduced by the filler (microscopic inhomogeneities) and chain entanglements due to the crosslinks [191]. An example of two-dimensional J_{CH}-resolved ^{13}C NMR spectra of cured, carbon black filled natural rubber, which displays the J_{CH} couplings in the F_1 dimension, and ^{13}C chemical shifts in the F_2 dimension, is shown in Fig. 31.

Since multiple-pulse decoupling is not applied, the J_{CH} values are not scaled; the J_{CH} values are 150 Hz for the C-2 carbon and 127 Hz for the carbons C-3, C-4

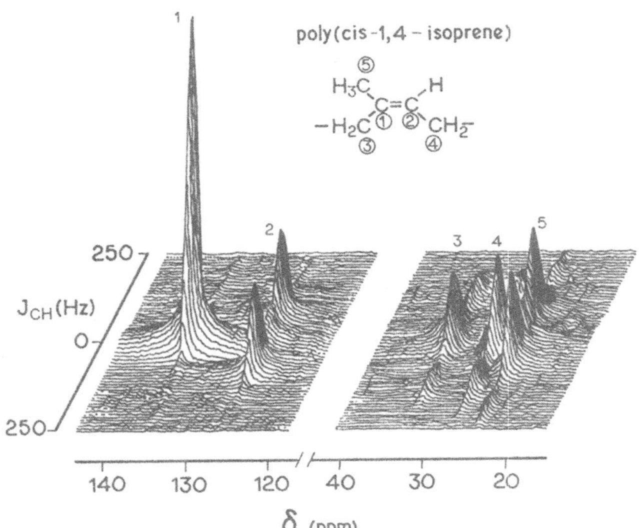

Fig. 31. Stacked plot of the heteronuclear two-dimensional J-resolved spectrum of cured, carbon black filled, natural rubber. The proton flip experiment was used with high-power proton decoupling during the detection time. The experiment was performed with the sample spinning at the magic angle (reprinted from Ref. [191] with permission)

and C-5. The values, which reflect the electronic environment and bonding characteristics, are identical with those found in a chloroform solution of natural rubber [191]. The investigated sample is prepared from natural rubber containing 50 phr (phr — parts per hundred of rubber) carbon black and 2.5 phr sulfur and the resonances of the crosslinking structural units (the average molecular weight between crosslinks is of the order of 10^4) are not observed; the sulfur content in the whole sample is 1.6 wt. % and the corresponding amount of carbons involved in each of possible mono- and oligo-sulfidic structures is probably in the order of 0.1 %.

The chemical structure of natural rubber and *cis*-polybutadiene crosslinked with dicumyl peroxide has been studied by both GHPD and CP techniques [195]. *Cis-trans* isomerization and the appearance of at least four structurally different methyl groups have been directly detected in natural rubber. The later results indicate that the simple combination of an allylic free radical is an oversimplification of the curing process. Crosslinked polybutadiene exhibits only methine and methylene carbons, with a small amount of methyl end-groups. In order to achieve better resolution of highly crosslinked vulcanizates, their increased molecular mobility (and hence the narrowing of the lines) is obtained by swelling in benzene to equilibrium. The difference between GHPD and CP spectra yields a broad resonance in the aliphatic region, indicating the presence of multiple chemical structures. Models for crosslinked networks are proposed from the results of NMR and Fourier transform infra-red (FTIR) investigations [193].

The new absorptions in the spectra of crosslinked rubber are assigned on the basis of ^{13}C solution NMR chemical shifts for a variety of model compounds, such as pentenes and mono-, di- and tri-sulfidic compounds, by using the ^{13}C chemical shift substituent effect. From the calculated values for particular structural units, the experimental spectra of a sulfur vulcanized natural rubber [194,195,106], natural rubber cured by accelerated sulfur vulcanization [197], γ-irradiation crosslinked natural rubber [198] and peroxide crosslinked natural rubber and *cis*-polybutadiene [193,199] are assigned.

By using this method, the chemical shifts of the resonances in the spectra of a sulfur vulcanized natural rubber (Fig. 32; expanded aliphatic region in shown in Fig. 33 [top]) are assigned to various units of the polymer network, which arise from structural modifications induced by the vulcanization [194,196,200]. Different sulfidic structures are found for unaccelerated and accelerated sulfur vulcanizations, respectively. With increasing amount of accelerator (as compared to the sulfur), the network structure exhibits less crosslinking, fewer main chain structural modifications, and fewer cyclic sulfide structures [197].

The GHPD technique has an advantage compared to cross-polarization in the more quantitative interpretation of absorption areas and line-widths. Furthermore, the cross-polarization dynamics for a spin system with lightly coupled protons, like in the elastomers is rather complex and at the present there is no corresponding theoretical explanation for this dynamical process. An example of both experiments performed on sulfur vulcanized natural rubber (10 % of sulfur) is shown in Fig. 32. The GHPD experiment is more sensitive to the mobile portions (uncrosslinked or lightly crosslinked samples), while the cross-polarization (CP) experiment is measuring the spectra of the more rigid or highly crosslinked parts of the network (the delay between GHPD pulse sequence is 5 s, and the contact time in CP experiment is

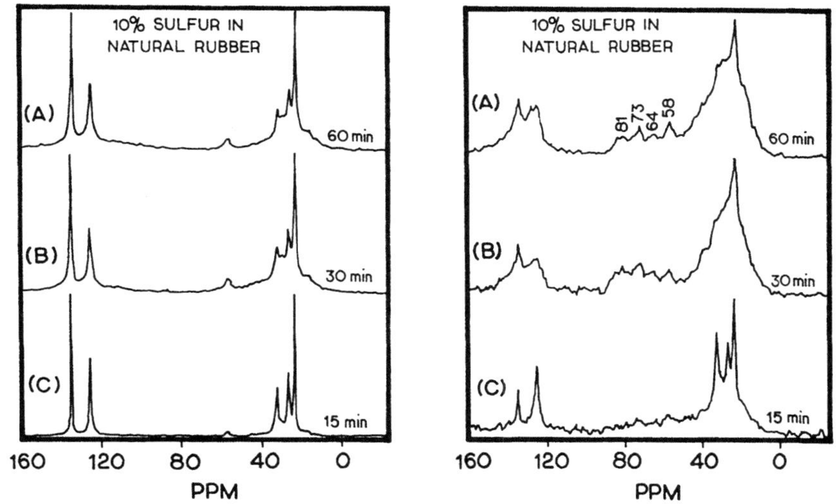

Fig. 32. GHPD (left) and cross-polarization (right) ^{13}C NMR spectra of natural rubber cured with 10% sulfur at 150 °C for 15, 30 and 60 minutes (adapted from Ref. [196])

1 ms). The lower signal-to-noise ratios of the CP spectra indicate that only a small portion of the sample is highly crosslinked. However, GHPD and CP spectra of rubber vulcanized with 30% sulfur for 60 minutes are very similar, indicating the similarity in the network structures of both rigid and mobile regions [196]. The observed differences in carbon spin-lattice relaxation times in the rotating frame determined by both, the direct and cross-polarization techniques, respectively, also indicate the selective sensitivity of the two techniques to the mobile and rigid parts of polymers, respectively [201].

A comparison between the peak intensities in GHPD and CP spectra (with the assumption that the resonances in both spectra have the same line widths) can be used as a qualitative measure for the local mobility of the particular structural units: higher values of the GHPD:CP intensity ratio indicate higher mobility. The mobility of the main chain carbons in acrylic-melamine coatings is found to depend on the extent of crosslinking and the type of crosslinking agent (fully alkylated melamines which also form melamine-melamine methylene crosslinks). However, the resonance areas of the side-chain acrylic carbons for a number of samples suggest their insensitivity to the copolymer composition, molecular weight and the glass transition temperature (T_g). In the cured and degraded copolymers a greater decrease in the mobility of main-chain carbons (lower GHPD vs. CP intensities) with degradation indicates the formation of acrylic crosslinks; analogous decrease for the melamine triazine ring indicates that the acrylic-melamine crosslinks formed by curing are hydrolized during the degradation and the melamine-melamine crosslinks, as was predicted from infrared spectroscopy, are formed [202,203]. This combination of GHPD and CP solid state NMR techniques is found to be very useful for monitoring the changes in the network structure during photodegradation, particularly in the case of acrylic coatings, where the degree of segmental mobility allows the successful simultaneous application of both techniques. An analogous study was made on the cure and degradation of acrylic-urethane coatings, where the mobility of the acrylic

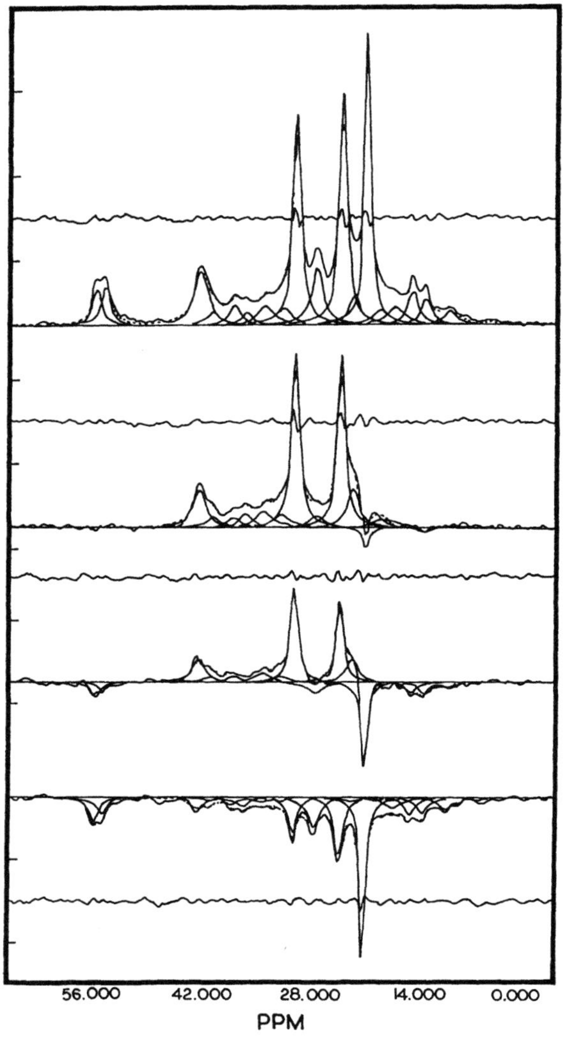

Fig. 33. ^{13}C NMR spectra of natural rubber cured with 10% of sulfur at 150 °C for 120 min obtained by the inversion recovery pulse sequence: $180° - \tau - 90° - t_d$. The τ values are 4 s (A), 0.2 s (B), 0.1 s (C), and 0.03 s (D) (from Ref. [200])

chains is greater (about 30%) than in the melamine coatings. It is found that the loss of urethane crosslinks is photochemical rather than hydrolytic in nature; the chemistry of crosslink degradation depends on the composition of the acrylic copolymers [203].

The line widths ($\Delta\nu$) of particular resonance lines in the GHPD experiment are related to the spin-spin relaxation time ($T_2 \sim 1/\Delta\nu$) and, hence, they reflect the change in molecular dynamics. Decreasing T_2 times, i.e. increasing of $\Delta\nu$ with the increasing crosslinking (larger amount of crosslinking agent or longer curing time) are generally observed for elastomeric networks [193,196,197]. However, besides the restriction in segmental mobility, the increased line broadening could also be a consequence of the chemical shift dispersion due to the formation of new crosslinked structural units. Generally, the major mechanisms of line broadening for ^{13}C resonances are those that i) produce a distribution of resonance frequencies (inhomogeneity in the static field B_0, variations in bulk magnet susceptibility within the sample, distribution of isotropic chemical shifts and residual chemical shift anisotropy), and ii) are related to the relaxation process (the natural linewidth and the residual static ^{13}C—^{1}H dipolar broadening from incomplete motional narrowing) [204]. The line widths and the line-broadening mechanisms for an epoxy resin (DGEBA) cured with piperidine are discussed in detail [205]. It should be noted that the linewidths of lightly crosslinked butyl rubber (0.27 and 0.17 crosslinks per 100 repeating units, respectively), as detected from the proton-decoupled solid state ^{13}C NMR spectra without magic angle spinning, are found to be unchanged with the introduction of crosslinks in the linear polymers [190]. The line widths in CP spectra of styrene-divinylbenzene networks are also approximately independent of crosslinking [206].

5.2.2 Spin-lattice Relaxation

The ^{13}C T_1 relaxation times in glassy (and particularly crystalline) polymers are substantially longer (approximately 10^0 to 10^3 s) as compared to the ^1H T_1. The low natural abundance of ^{13}C nuclei eliminates the spin-diffusion process, which yields a single T_1 for all the coupled protons in the sample. Hence, each polymer carbon will have its own relaxation time, influenced primarily by the molecular dynamics. ^{13}C spin-lattice relaxation times reflect motional frequencies in the 15 to 75 MHz range. The observation of individual carbon T_1s in the solid state has an advantage in distinguishing between polymer side-groups dynamics and the relaxation of carbons in the main chain, thus yielding additional information on polymer dynamics. The pulse sequence most frequently used for T_1 measurements in solid state is the standard inversion recovery method [174] modified with high power decoupling during the acquisition. A direct application of a 90° pulse on carbons is suitable for the study of elastomers, because of the relatively short ^{13}C T_1 relaxation times; for more rigid polymers a building-up of the carbon magnetization via cross-polarization is preferred.

An early solid state ^{13}C NMR study (without magic angle spinning) of butyl rubber showed that the carbon T_1 relaxation times at room temperature are not affected by the small degree of crosslinking (crosslink density up to 4.5×10^{-5} mole/cm^3), thus indicating that crosslink levels similar to those in commercial vulcanizates do not affect relatively rapid segmental motions which determine T_1 relaxation times [190]. A similar insensitivity of carbon T_1s at room temperature is found in vulcanized natural rubber with significantly higher degree of crosslinking [196,197].

However, the study of ^{13}C T_1 relaxation time over a wider temperature range, combined with the application of computer decomposition of overlapping spectral regions, shows the effect of crosslinking on each detectable and resolvable carbon in the vulcanized natural rubber (10% of sulfur). The spectra of crosslinked polymers generally consist of overlapping signals, and a decomposition of complex spectra is required in order to isolate the peak intensities of particular signals which, although occasionally separated by relatively small differences in chemical shifts, have different relaxation times. An example of the decomposition of overlapping aliphatic region in vulcanized rubber as a function of the increasing τ in the inversion recovery pulse sequence is shown in Fig. 34.

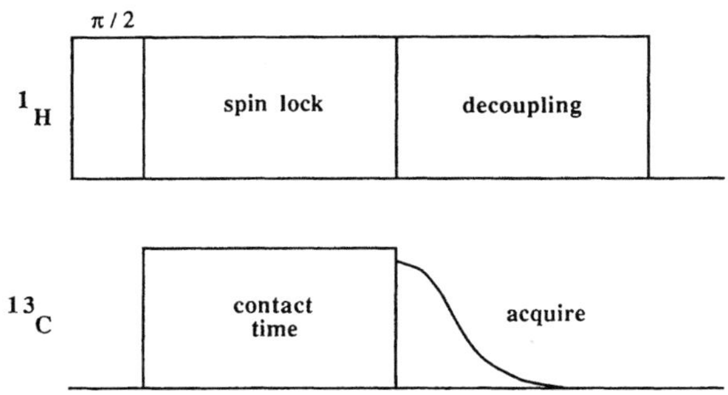

Fig. 34. Cross-polarization (CP) pulse sequence

The temperature dependence of T_1 for each polyisoprene carbon exhibits the expected minimum in the curve at room temperature (at 37.7 MHz); its position and magnitude, accompanied by the broadening, is generally shifted to higher temperatures with increasing degree of crosslinking. The methyl group shows more sensitivity to motional restriction due to the crosslinking than the backbone carbons [207]. A quantitative analysis of vulcanizate GHPD spectra, containing some signals with relatively longer T_1s (longer than 1/5 of the delay time) is done by using the area correction for the non-fully relaxed nuclei [200].

5.2.3 Cross-Polarization

Owing to the relatively long ^{13}C T_1 relaxation times (for example ^{13}C $T_1 = 5200$ s for the crystalline component of polyethylene T_1) and their low natural abundance, GHPD and related techniques, based on application of 90° pulse on carbon atoms, are not generally suitable for glassy and crystalline polymers. In order to increase the sensitivity (signal-to-noise ratio) of the dilute spins (S) it is more advantageous to build up the carbon magnetization via polarization transfer from the abundant proton spin system (I) by application of two resonant rf fields B_1^I and B_1^S, respectively, which satisfy the Hartmann-Hahn condition [208]:

$$\omega_{1I} = \omega_{1S} = \gamma_I B_{1I} = \gamma_S B_{1S} \tag{48}$$

The cross-polarization (CP), i.e. the transfer of I-spin polarization to the dilute spins (S), is a double resonance experiment in which the I and S spins are coupled by a certain interaction, determined by the cross relaxation time T_{IS}. From the dynamics of the CP process, usually described with the spin temperature concept, the following equation for the time dependence of S-spin polarization could be derived:

$$\frac{dS}{dt} = \frac{S_0 \exp(-t/T_{1\varrho}^I) - S}{T_{IS}} - \frac{S}{T_{1\varrho}^S} \tag{49}$$

where S_0 is maximum S-spins polarization with no dissipative process, $T_{1\varrho}^I$ and $T_{1\varrho}^S$ are rotating frame relaxation times for spins I and S, respectively, and T_{IS} is the cross-polarization time under spin-locking conditions. In the case of protons (or fluorine) as abundant spins I and carbons (or silicon) as dilute spins S, the relations $T^S_{1\varrho} \gg T_{IS}$ and $T^S_{1\varrho} > T^H_{1\varrho} \gg T_{IS}$ are generally valid. With these assumption Eq. (49) can be written as:

$$S \sim S_0 \left(1 - \frac{T_{CH}}{T_{1\varrho H}}\right)^{-1} \left[\exp\left(-\frac{t}{T_{1\varrho H}}\right) - \exp\left(-\frac{t}{T_{CH}}\right)\right] \tag{50}$$

where subscripts C and H stand for carbon and proton, respectively. The cross-polarization pulse sequence consists of: a) establishing ^1H spin polarization in B_0; 2) cooling of the proton spin system, which is most frequently done by spin-locking in the rotating frame B_1 [209]. 3) the bringing ^{13}C and ^1H spins in contact during a certain period of time by the Hartmann-Hahn match; 4) turning off B_{1C} and recording ^{13}C free induction decay (FID); 5) repeating the pulse sequence following the delay time t_d (Fig. 34).

Building up the carbon magnetization by the CP process has two major advantages: a) the relaxation process is much more rapid, because it is determined by the proton T_1, as compared to the carbon T_1 (which allows relatively short recycle delay t_d between the pulse sequences), and b) intensity enhancement is four times greater than in the GHPD experiment ($\gamma_H/\gamma_C = 4$). In an early CP MAS study of γ-irradiation (6.24 MGy) crosslinked polyethylene (at 15 MHz) only a single low-intensity resonance (about 1%), tentatively attributed to the chain ends or to the vicinity of crosslinks was observed [210]. Because of the relatively small fraction of carbons associated with the macromolecular network structure, one requires the enhanced signal to noise ratio, accompanied by stronger magnetic fields or enriched ^{13}C samples in order to detect the network absorptions. The effect of the magnetic field (resonance frequencies of 15, 25 and 50 MHz) on the spectra of urea-formaldehyde resins and related monomers indicate the important advantage of better resolution. The higher field also suppresses the additional line broadening due to the presence of ^{14}N—^{13}C coupling in this particular polymer, and in nitrogen containing polymers in general [211]. However, a study of cured norbornene end-capped polyimides at 15, 37.7 and 75.5 MHz, respectively, indicate that the increase in the static field by a factor of five has no significant effect on the resolution, although a significant improvement in sensitivity is observed [212]. The application of higher magnetic fields usually causes the appearance of spinning sidebands, which are generally undesirable in the assignment of spectra, although their presence could yield additional information about the spin system under the CP MAS conditions. The relative amplitude of the sidebands is a function of the anisotropy in the chemical shift tensors $\boldsymbol{\sigma}$, and they can be removed if the magic angle spinning frequency is equal or greater than $\Delta\sigma$, the difference between the minimum and maximum chemical shift principal values. There is a pulse sequence for sidebands suppression, which is frequently used in qualitative and — if there are no artificial effects on signal intensities and line widths — also in the quantitative ^{13}C NMR studies [213].

Characterization and assignment of particular signals in glassy insoluble polymer networks is one of the major analytical applications of cross-polarization and magic angle spinning techniques. The chemical shift assignments are usually made by a) comparison with the chemical shifts of analogous model compounds in the solution,

b) calculating the ^{13}C chemical shift substituent effect and sometimes c) by the application of dipolar dephasing (DD) pulse sequence in order to distinguish between CH, CH_2 and CH_3 groups (see Sect. 5.2.5). The structural characterization for a number of crosslinked polymers has been made with the CP MAS technique: cured acetylene-terminated resins [214, 215], cured phenolic resins [211, 216, 217] (in combination with ^{15}N NMR) [218], cured epoxy resins [219-224], cured norbornene end-capped polyimides [212], γ-irradiation crosslinked ethylene-propylene copolymers [225-228], plasma polymerized hydrocarbons [229, 330], and toluene [231], styrene crosslinked unsaturated polyesters [232, 233], cured furfuryl alcohol resins [234, 235], cured urea-formaldehyde resins [211], cured melamine formaldehyde- and urethane-acrylic coatings [202, 203, 236, 237], γ-irradiated peroxide cured polyethylene [215] and γ-irradiation crosslinked polybutadiene [238].

An example for the polymer network characterization by the ^{13}C CP MAS NMR is shown in Fig. 35. The chemical structure of the cured polystyrylpyridine resins (PSP), synthesized from terephthalic aldehyde and collidine (2,4,6-trimethylpyridine), is determined from CP-MAS spectra by comparison with the solution state spectra of the model compounds and supported by selective DD observations. The CH and CH_2 of the crosslinking points, as deduced from the model BP_2, give rise to a composite line at about 45 ppm; the assignment of other signals is indicated in the figure [239].

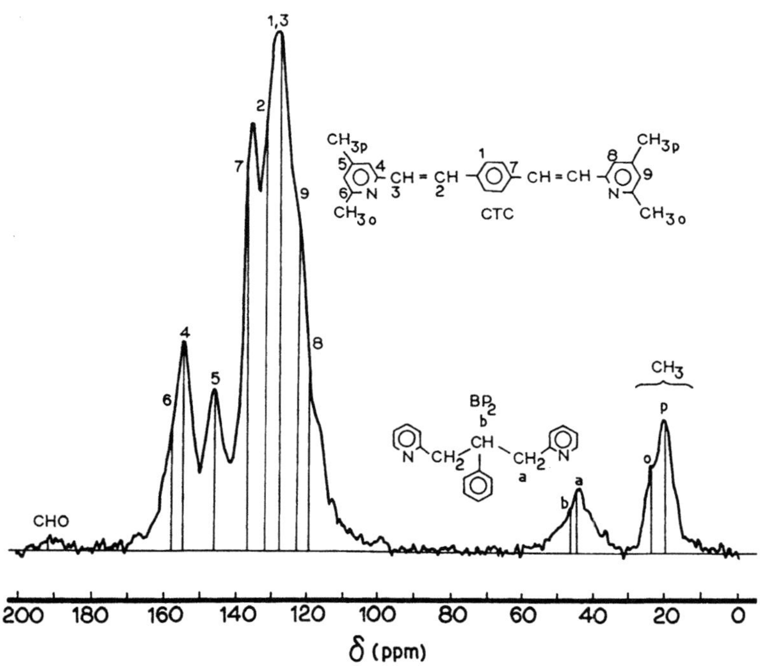

Fig. 35. ^{13}C CP MAS NMR spectra and peak assignments of a cured polystyrylpyridine resin. Stick spectra are drawn from ^{13}C NMR spectra of the model compounds CTC and BP_2 in solution (reprinted from Ref. [239] with permission)

A complementary study of the curing of norbornene end-capped polyimides by the solution and solid state NMR, respectively, shows that the same polymerization mechanisms are obtained from both techniques: a retro-Diels-Alder reaction of the end-capped nadic group and further Diels-Alder reaction of the oligomers and cyclopentadiene. The various end-capped groups polymerize through reaction of the nadic olefinic group to form the crosslinked polymer. However, crosslinked samples are only partially soluble, and all the resonances observed in the CP MAS spectra are not detected in the corresponding solution spectra. Some of them appear in the solution spectra at higher degree of crosslinking, although their presence in the solid state spectra is confirmed in the earlier stage of the curing process [212)].

The distribution of styrene crosslinks in cured unsaturated polyesters based on maleic anhydride, phthalic anhydride, propylene glycol and dipropylene glycol is found to be dependent on the molar ratio of styrene vs. maleic/fumaric unit. If the molar ratio is higher than 1, the dyad distribution is predominant; lower molar ratios yield mostly styrene microdomains, i.e. n-ads with more than 2 styrene repeating units [232)].

The spectra of *N*,*N*′-dimethylolurea formaldehyde resins cured for different times are accurately simulated by using a set of nine different chemical shift positions, corresponding to different structural units in a linear and crosslinked systems. The changes in the intensities of CP MAS signals (obtained at 1 ms contact time) with the increasing curing time indicate the disappearance of methoxymethylol and methylol groups on the secondary and tertiary nitrogen atom, and the growth of methylene bridges between secondary nitrogens and tertiary and secondary nitrogens as the curing time increases [211)].

Cured phenol formaldehyde resins, prepared under alkaline conditions, exhibit three methylene signals in the aliphatic region. They are assigned to free paraformaldehyde, methylol groups and to bridging methylenes, respectively. Greater observed line widths (150 to 250 Hz), as compared to the glassy polymers (about 60 Hz), seem to be a general characteristic of crosslinked two-component resins because of a) a variety of closely related isomeric forms and b) a residual chemical shift dispersion due to the variation of the spatial arrangement of the surrounding groups [216)]. Thermal degradation of these resins (prepared with ^{13}C-enriched formaldehyde) was followed in both, air and vacuum [240, 241)].

Investigation of the furfuryl alcohol resins structure during the crosslinking process show that in the very early stages there is no evidence for the existence of appreciable concentrations of formaldehyde, methylol groups, or dimethylene ether linkages. Major crosslinking processes involve linkages with bridging CH_2 groups, rather than substitution at the 3- and 4-positions of furan rings [235)]. Spectra of phenol-formaldehyde resins (of resol type) indicate a large structural diversity of the samples prepared in various ways and with different molar ratios of formaldehyde to phenol. The direct involvements of the hydroxyl group of the phenol ring and condensation of methylene bridges with hydroxyl groups during the curing process is detected, as well as the presence of $-CH_3$ and $-CHO$ groups directly attached to the phenol rings [242)].

In the case of epoxy resins (DGEBA cured with dimethylbenzylamine) a decrease in the intensities of the methylene epoxide ring carbons accompanied with an increase in the intensity of the methylene carbons attached to the oxygen atom is observed.

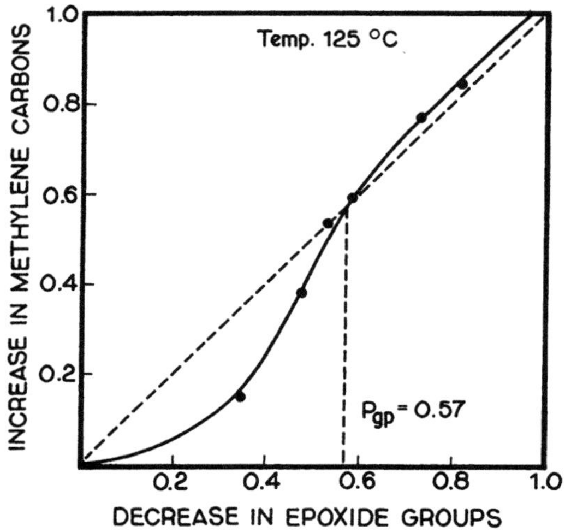

Fig. 36. Plot of increase in methylene resonance against the decrease in the epoxide group (reprinted from Ref. [221]) with permission)

A plot of these intensities, oxymethylene group increase vs. epoxide group decrease, is shown in Fig. 36. The conversion of one epoxide group should yield under ideal conditions one oxymethylene unit. However, the experimental data show a deviation from ideality until the extent of the reaction reaches 0.57. Nonstoichiometric behaviour in this range could be a consequence of a) side reactions which do not result in a formation of oxymethylene units (corresponding resonances are not observed by NMR and the amount of possible by-products is smaller than 2 to 3%), and b) the changes of molecular motion at different stages of curing which may have an effect on the cross-polarization, i.e. at the early stage of curing the efficiency of the polarization transfer to the oxymethylene carbons may be lower when compared to the final stage of curing. The transition at the extent of reaction 0.57 may suggest the gelation point, from which the functionality of the epoxide (2.6) is determined [221)].

The molecular motions in the piperidine cured epoxy resins are investigated over a wide temperature range by assuming two models for the distribution of correlation times: homogeneous (all processes are described by the same nonexponential autocorrelation function) and inhomogeneous (there is a distribution of the single-relaxation processes), respectively [220)]. It is found that the analysis of CP NMR line shapes is extremely sensitive to the chosen model. Although the calculated energies of activation are consistent with the mechanical spectroscopic data, the 180° flipping of the phenylene ring (as calculated from the two models) is either approximately 3000 slower (inhomogeneous distribution) or 20 times slower (homogeneous distribution) than that for the motion that accounts for mechanical loss. However, the same correlation time as in the mechanical relaxation processes could be obtained by assuming the smaller diffusive steps for the phenylene rings, such as 3° (inhomogeneous distribution) or 40° (homogeneous distribution), rather than large reorientational jumps [220)].

A quantitative analysis of the absorption area in the CP spectra, which is important in the analytical and kinetical studies, is possible only if the degree of polarization transfer from protons to each carbon atom is constant. The intensity of each carbon signal, S(t), as follows from Eq. (50), depends on its particular T_{CH} (or $T_{CH(SL)}$, cross polarization transfer under spin-locked conditions; $T_{CH(ADRF)}$ denotes CP under adiabatic demagnetization in the rotating frame) and $T_{1\varrho H}$; consequently, the exponential growth of carbon magnetization reaches a maximum S at different contact times for different atoms. Quarternary carbons (proton starved) and methyl groups (owing to its mobility which averages static dipole interactions involved in cross-polarization) usually exhibit a decreased peak intensity at shorter contact times, as compared to the CH_2 and CH signals [7]. However, it is observed that some polymers have the same signal intensities in CP spectra as in fully relaxed GHPD spectra. Thus, by choosing the one appropriate contact time it is possible to make some quantitative analysis for particular samples. An additional difficulty for crosslinked polymers arises from usually overlapped, and sometimes unresolvable carbon lines which have different T_{CH}. Furthermore, the networks are also heterogeneous materials and $T_{1\varrho H}$ is not necessarily described by a single relaxation time.

From quantitatively reliable spectra (obtained under the appropriate conditions), the curing process of solid phenolic resins is studied by the ratio of methylene/methylol peaks (crosslinking by the conversion of $-CH_2OH$ to $-CH_2-$ groups) as a function of the curing time and curing temperature [243]. The degree of cure for phenolic resins is also followed via the ratio of $-CH_2-$ (methylene bridges between aromatics) vs. Ar_1 (aromatic carbon carrying the hydroxyl group) peak areas. A 2 ms contact time and 8 s delay time between transients are found to be the optimum conditions for a resin cured for 2 h at 70 °C (i.e. the largest peak areas for aromatic carbons are obtained). Under these conditions the same value of CH_2/Ar_1 ratios are measured (about 0.64) as in the case of fully relaxed GHPD spectra with 120 s delay between pulse sequences [244].

Although quantitative areas for all absorptions can not be obtained from a single spectrum, an assumption that a $-C{=}O$ group adjacent to a $C{=}C$ bond has similar relaxation times, and hence they have comparable peak, area is made. Based on this assumption the conversion of various multifunctional monomers, which yield highly crosslinked polymers with methyl methacrylate, is determined from the spectra recorded at 1 ms contact time. It is also found that the conversion of these comonomers increases with the increasing length and flexibility of the spacer between the methacrylate double bonds [245].

The spectra of acetylene-terminated polyimide resin, cured and postcured polyimide polymer recorded with 1 and 2 ms contact time, respectively, show no differences, indicating that 1 ms is sufficient to produce the full intensity for all types of carbons. Consequently, the areas under the particular peaks are a quantitative measure of the fraction of the types of carbons. Figure 37 shows the difference spectrum obtained by subtraction of the spectrum of the uncured resin from that of the cured resin. The results indicate that the major crosslinking mechanism involves loss of the acetylenic end-capping groups (84 ppm). It is estimated that no more than 30 % of the acetylene end groups trimerize to benzene during crosslinking (increase in intensity of the carbon resonances in the regions of 125 to 129 and

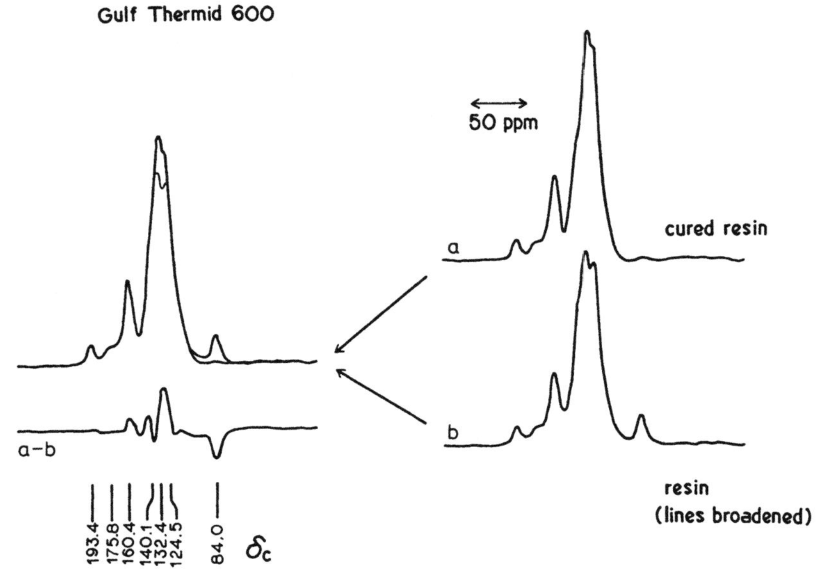

Fig. 37. Difference spectrum (a–b) obtained by subtraction of the uncured polyimide resin spectrum (with lines artifically broadened), b) from that of the cured polyimide polymer (a). Small deviations in the difference spectrum should be ignored since line broadening across the cured polymer may not be uniform (reprinted from Ref. [214] with permission)

140 to 144 ppm). The remainder appears to be consumed by addition reactions, i.e. conversion of acetylenic groups to vinyl carbons (125 to 130 ppm); three types of reactions are suggested: 1) addition to a carboxylic acid functional group, 2) Friedel-Crafts type aromatic alkenylation and 3) addition at an amide which, by inference, should be associated with the observed carboxylic acid [214]. A study of γ-irradiation induced crosspolymerization (non random crosslinking) of a model acetylenic macromonomer poly(1,11-dodecadiyne) shows that bridges between chains are of the acetylenic structure ($-C{\equiv}C-$) instead of the possible butatriene ($=C=C=$) [246].

Gamma-irradiated *cis*-1,4-polybutadienes exhibit a decrease in intensity of the $-CH=$ signal, indicating loss of double bonds due to the crosslinking, as well as the appearance of new overlapping absorptions attributed to the CH groups in H-crosslinks and $-CH_2-$ groups adjacent to the crosslinks. From the quantitative peak intensities, obtained by computer simulation, the yields of crosslinking, G(X), and loss of double bonds, G—(C=C), are found to be about 23 for the sample irradiated with 0.44 MGy, and 14 for the sample irradiated with a 3 MGy dose. G-values calculated from the NMR method are higher than the values obtained from other methods (swelling, elastic modulus or sol fraction analysis), indicating that certain crosslinks are formed in clusters. The apparent discrepancy arises because NMR technique determines total crosslink yield, including both clustered and intramolecular crosslinks, while the usual methods are sensitive only to isolated crosslinks [238].

The γ-irradiation induced crosslinking in ethylene-propylene rubber (EPR) was investigated qualitatively and quantitatively by the CP technique [225-227]. New peaks in the irradiated copolymer are ascribed to a) crosslinks at sites of the lost methyl group, b) crosslinks at the lost hydrogens of methylene in either ethylene or propylene units and c) short branches (high field methyl groups at the end of the short branch are detected). It is found that the molar ratio of ethylene and propylene in EPR estimated from the relative intensities of peaks is reliable under appropriate experimental conditions, i.e. 10 ms contact time and 10 s delay time between transients. Quantitative analysis of the EPR irradiated with 0.64 MGy yields 3% of crosslinks of the (a)- and (b)-type, respectively, and 5% of short branches. Since the NMR technique detects the crosslinks in a microstructure (C—C bonds introduced into the EPR main chain, up to about three chemical bonds), regardless of whether the side chains are open at the other end (branches) or not, the fraction of crosslinks effective for gel-formation should be corrected by subtracting the amount of branches formed, which have no effect on gel formation. Thus, the final result of 1% of crosslinks is found to be in a fair agreement with the value of 0.53% obtained from a Charlesby-Pinner plot [227]. EPR rubbers irradiated with a large dose (10 MGy) show both, a new peak attributed to methyl groups resulting from chain scission (14.3 ppm) and increased intensity in the region of. H-crosslinks (39 to 47 ppm). Quantitative analysis from the spectra recorded at 1 ms contact time and 5 s recycle time yields values G(S) and G(X) for scission and crosslinking, respectively, which are in good agreement with the values obtained by sol extraction, indicating that crosslinks are apparently not formed in clusters, as has been suggested for irradiated semicrystalline polyethylene [228].

In a study of the curing and degradation of melamine formaldehyde-acrylic copolymers a relatively long contact time of 5 ms is chosen to obtain representative spectral populations (i.e. minimize intensity distortions in spin counting) even for those nuclei with small static $^1H—^{13}C$ dipolar interactions. ^{13}C NMR integrated intensities, corrected for the estimated spinning sideband contributions (arising from a sample spinning at 5 kHz) and calculated results, assuming complete primary crosslinking and no side reactions (melamine self-condensation reactions), are found to be in qualitative agreement; the experimental values are in the range from 61 to 154% of the calculated intensities. More accurate quantitative information could be obtained with a quantitative calculation of spinning sideband intensities and spectral simulation [236]. However, at the slower spinning frequencies (about 3.5 kHz) the contribution from spinning sidebands is rather small (less than 10%) except for the styrene, and possibly for the carbonyl resonance. For all the samples only three of nine resonances have averaged resonance areas (measured without the correction for spinning sidebands) significantly different from that predicted from stoichiometrical calculations (70, 84 and 134%, respectively); the lower values could result from neglecting the sidebands [202].

Highly crosslinked networks based on plasma polymerized hydrocarbons (between 1:5 and 1:3 crosslink and/or cyclic structure per carbon atoms) show no change in relative spectral intensities to contact times from 0.8 to at least 3 ms. In addition, spectra show no noticeable distortions with sample aging even though the ^1H $T_{1\varrho}$ changes significantly, suggesting the recombination of free radicals originally created by the plasma. The $T_{1\varrho}$ relaxation curve is also found to be a sum of at least two components which are attributed to the protons near paramagnetic sites (short, 100–200 μs relaxing component) and protons spatially removed from the influence of the paramagnetic sites (a 7 ms slowly relaxing component) [230].

Plasma polymerized toluenes exhibit only small differences in relative peak areas in the contact time range from 0.5 to 5 ms, suggesting that the integration of the peak areas will result in complete spin counting. By using the specific ^{13}C labeling at the toluene methyl (Fig. 38b) and C-1 ring carbon (Fig. 38c) it is possible to trace reaction pathways of labeled toluenes in the plasma process. Spectra are resolved into four bands: protonated and non-protonated unsaturated carbons, and methyl and other saturated aliphatic carbons; assignment is supported with the dipolar dephasing experiment, which differentiates between the protonated and non-protonated carbons (see Sect. 5.2.5). Increased peak intensities in the spectra of the labeled toluene, compared to those of natural abundance (Fig. 38a), are a consequence of the incorporation of a) methyl carbons primarily into saturated structures and b) phenyl C-1 carbons primarily into unsaturated structures in the polymer. From relative intensities (obtained from the linear plot of particular component

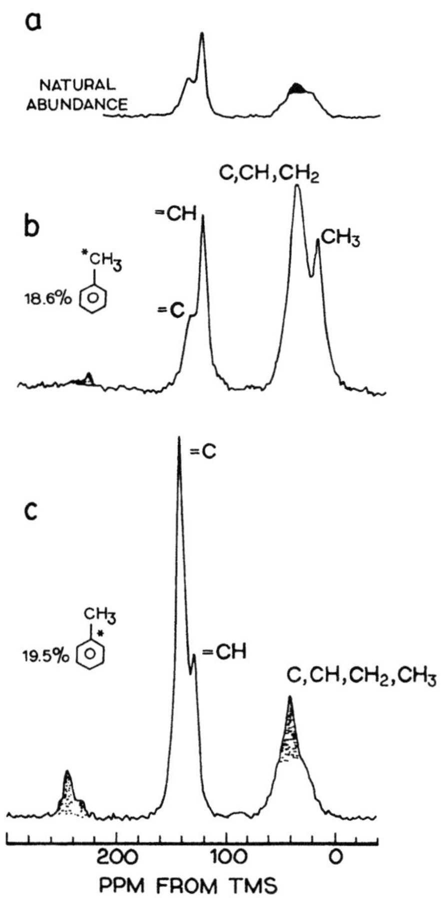

Fig. 38a–c. CP MAS ^{13}C NMR spectra of plasma polymerized toluene with natural ^{13}C abundance **a**, ^{13}C-methyl enriched **b** and phenyl ^{13}C-1 enriched toluene **c**. Shaded peaks are spinning sidebands (adapted from Ref. [231])

signal as a function of enrichment level) it is possible to determine the polymer sites into which specific toluene carbons are incorporated. For example, about 20% of toluene C-1 carbons become saturated, while 24% of the methyl carbons end up as unsaturated carbons during the plasma polymerization. Furthermore, approx. 30% of the toluene double bonds saturate during polymerization and about 20% of ring hydrogens are displaced by carbons. However, the observed ^{13}C resonances, assigned to benzyl and tolyl structures, could actually involve a number of different structures with a range of chemical shifts, such as substituted cyclohexenes and cyclohexadienes. Based on this assumption and the quantitative analysis, a structure of the crosslinked plasma polymerized toluene was proposed (Fig. 39) [231].

Quantitative analysis of the curing process of polystyrilpyridine resins is made by following the dynamics of the cross-polarization process over a wide range of contact time. Intensities of the particular signals (see Fig. 35) as a function of the contact time are shown in Fig. 40. The long-time magnetization decays for each carbon are exponential and described by the same $T_{1\varrho H}$, thus indicating a homogeneus network structure in terms of the NMR scale. The maximum carbon polarization in the absence of dissipative process ($T_{1\varrho H}$), S_0, is calculated by using modified Eq. (45). The number of pyridine rings (which is not modified by the chemical reactions involved) is taken as an internal reference; the number of methyl (*ortho-* and *para-*), aldehyde and aliphatic crosslinking groups per 100 pyridine rings is calculated from S_0 intensities as presented in Table 4. From the presented data, it is found that the relative numbers of various functionalities (D), i.e. variations of x from one curing stage to the next, follow the relationship $-D_{CH_3} = -D_{CHO} + D_{cl}$. Thus, it can be concluded that the loss of methyl groups is mainly obtained via the two mechanisms, polycondensation and crosslinking. Decomposition of the methyl carbon line in two separate signals at 20 and 24 ppm (*p-* and *o-*methyl carbons, respectively), makes possible the observation of different reactivities of *o-* and *p-*methyl groups in collidine. The *p-*Methyl reactions occur only when most of the *o-*methyl functions have disappeared [239].

The initial building-up of the carbon magnetization via polarization transfer is characterized with the spin-lock cross-polarization time T_{CH}. It is sensitive to the static interactions, i.e. transfer of the polarization via static dipolar interactions (proportional to e_{CH}^{-6}). Hence, the shortest T_{CH} will have carbons a) in more rigid systems and b) with more directly attached protons, particularly the CH_2 (or

Table 4. Number of methyl (x_{oCH_3} and x_{pCH_3}), aldehyde (x_{CHO}) and crosslinking (x_{cl}) groups per 100 pyridine rings in cured PSP[a]

Curing conditions 8 h at 200 °C +	x_{oCH_3}	x_{pCH_3}	x_{CHO}	x_{cl}
0 h	35	58	40	26
2 h at 250 °C	22	61	29	32
16 h at 250 °C	18	59	27	34
48 h at 250 °C	19	48	25	41

[a] Adapted from Ref. [239]

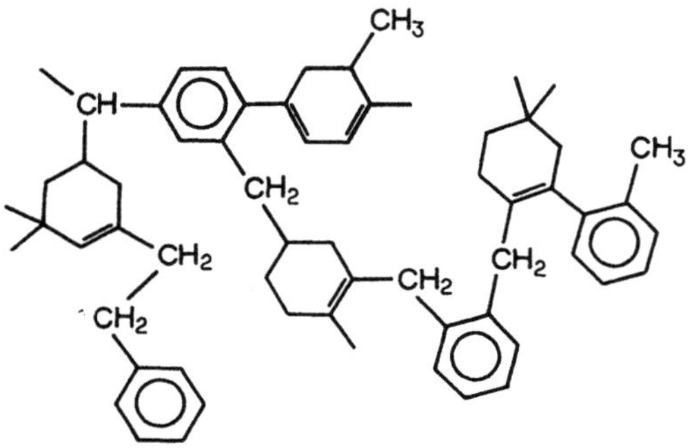

Fig. 39. Proposed structure of the plasma-polymerized toluene, based on the data from solid state ^{13}C NMR of isotopically labeled toluene (from Ref. [231])

static CH_3) groups. Generally longer T_{CH} values are observed in poly[oligo(ethylene glycol)dimethacrylate]s with larger number (14) of ethylene glycol units between crosslinks (rubbery gel), than in the highly crosslinked glassy homologues [247]; dependence of T_{CH} on the extent of cure for tetra(ethylene glycol)dimethacrylate is found to be relatively unaffected above the vitrification point (i.e. at about 60% conversion) [248].

A decrease of the signal intensity in the CP experiments due to the ^1H $T_{1\varrho}$ relaxation time provides no information about molecular dynamics. The proton relaxation in its rotating field is dominated by the spin-diffusion process, resulting in a single proton $T_{1\varrho}$ for a whole sample, if it is homogeneous and the protons are strongly coupled. A constant value for ^1H $T_{1\varrho}$ (within an experimental error) indicates dynamical homogeneity. This value is also observed in a piperidine cured epoxy resin — crosslinked poly[oligo(ethylene glycol)dimethacrylate]s) (the ^1H $T_{1\varrho}$ calculated from Eq. (49) is in agreement with the direct observation from proton

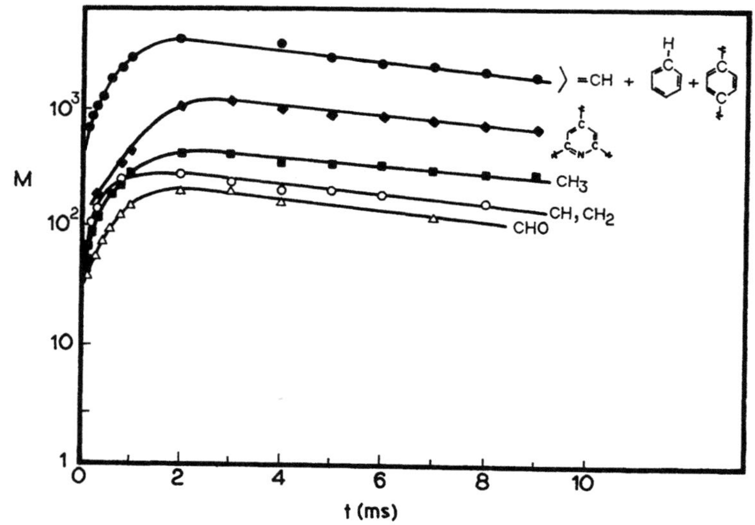

Fig. 40. Evolution of carbon magnetizations in a cured polystyrylpyridine as a function of the contact time t (reprinted from Ref. [239] with permission)

pulsed NMR)[188]. An increase in $^{1}H\ T_{1\varrho}$ with the increasing number of ethylene glycol units between crosslinks is observed[247] with cured tetra(ethylene glycol)dimethacrylate. The $^{1}H\ T_{1\varrho}$ values do not change significantly with the increase in crosslinking[248].

A combination of both high resolution techniques, in the solution and solid state, respectively, is applied for the "in situ" study of the curing of an epoxy resin (DGEBA with triethylenetetramine). The development of the network is followed by performing the reaction in the rotor and switching from the solution NMR experimental conditions to the cross-polarization and magic angle spinning[249]. With increasing crosslinking the signals in the solution spectra become broader and less intense (Fig. 41 a–c). However, in the vicinity of the gel point it is not possible to obtain a spectrum by either cross-polarization or single-pulse excitation (GHPD).

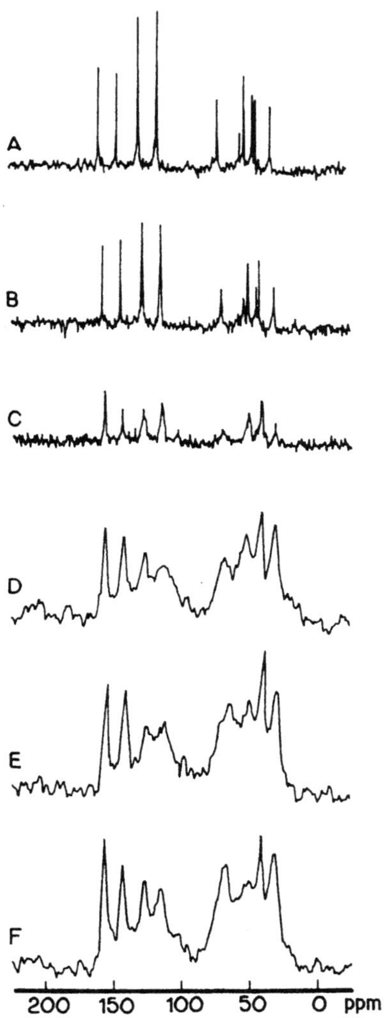

Fig. 41. ^{13}C NMR spectra recorded during the crosslinking of an epoxy resin: solution conditions (A–C; before the gel point) and CP MAS in the solid state (D–F; after the gel point). The reaction times are 5 to 25 min (A), 26 to 46 min (B), 46 to 66 min (C), 123 to 133 min (D), 135 to 145 min (E) and 350 to 380 min (F) (adapted from Ref. [249])

At this stage of the curing process there is significant molecular motion in the mid-kiloherz region that is similar to the magnitude of the stochastic averaging of dipolar couplings due to the random molecular motions (decoupling field γB_1 of approximately 48 kHz). With further crosslinking, the amplitudes of the molecular motions decrease and the spectra are detected under cross-polarization conditions (Fig. 41 d–f); the spectra show an increase of the signal intensity in the 60 to 75 ppm region, i.e. the formation of secondary alcohol groups as a result of epoxide ring opening during condensation. However, these spectra are not suitable for quantitative analysis because of the low signal-to-noise ratio caused by the low ^{13}C natural abundance and the short time between reaction steps available for data collection.

Maximum polarization transfer in solids is achieved when ω_{1C} and ω_{1H} are equal. However, it is found that polybutadiene (as well as adamantane) exhibits maximum transfer rates (higher signal intensities) when the carbon spin locking radio frequency is offset from the exact Hartmann-Hahn match (Eq. (48)) by integrals of the spinning frequency. High-speed magic angle spinning is shown to amplitude modulate (AM) the static $^{13}C—^{1}H$ dipolar interaction and frequency modulate (FM) the homonuclear proton dipolar decoupling in mobile solids. Under a mismatch of ω_{1C} and ω_{1H} (held constant at 12 kHz) for a multiple of the MAS frequency a sideband pattern, which changes with the increased homonuclear proton coupling (degree of crosslinking), is observed (Fig. 42). The FM part of the pattern (whose intensities are determined by a series of Bessel functions of the first kind)

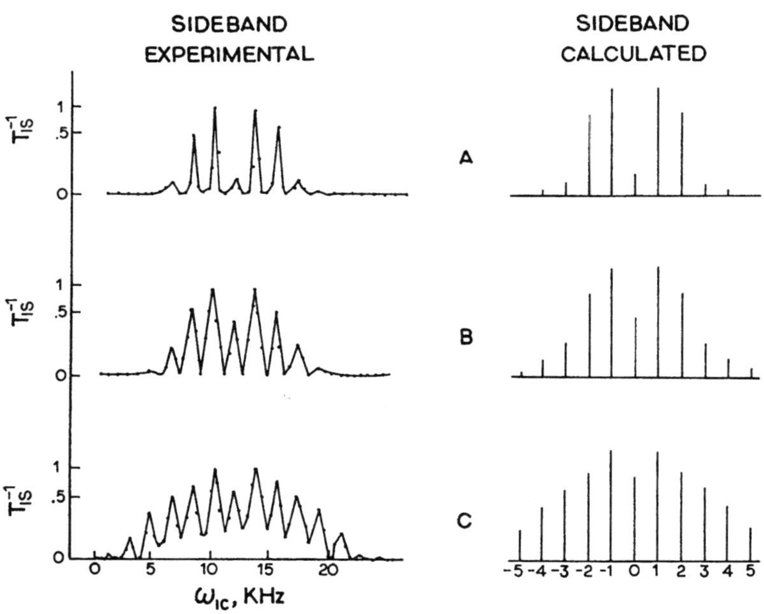

Fig. 42. Comparison of the experimental and calculated Hartmann-Hahn mismatch sideband patterns for polybutadienes with the average molecular weight between crosslinks M_c = 6810 (A), M_c = 2730 (B) and M_c = 1020 (adapted from Ref. [250])

could be used to characterize the physical mobility of the crosslinked rubber. The crosslink-density can be related to both the FM index of the Bessel functions and the fraction of FM character of the pattern. By using this technique it is possible to follow the effect of crosslinking on the low-frequency motions (kHz range) in elastomers [250].

5.2.4 Spin-lattice Relaxation in the Rotating Frame

The ^{13}C $T_{1\varrho}$ relaxation times, like the ^{13}C T_1 times, but unlike the corresponding ^{1}H $T_{1\varrho}$ times, are not influenced by the spin-diffusion, and therefore the relaxation of each carbon can be followed. In the $T_{1\varrho}$ pulse sequence a carbon magnetization is built-up via cross-polarization, and the carbons are held in the rotating frame without direct contact with the proton reservoir for a variable period τ, allowing the ^{13}C polarization to decay in its own rotating field (Fig. 43). This structure of the pulse sequence results in a sensitivity of $T_{1\varrho}$ to molecular motion in the 10 to 100 kHz frequency range, characteristic for the relatively long-range cooperative motions of a polymer chain below the glass transition. The Hamiltonian governing the proton-carbon (I—S) spin system under the Hartman-Hahn condition ($\omega_{1I} = \omega_{1S}$) can be expressed by the following equation:

$$\mathbf{H} = -\omega_1 \mathbf{I}_z - \omega_1 \mathbf{S}_z - \frac{1}{2}\mathbf{H}_{II}^{(0)} + \mathbf{H}_{II}^{(ns)} + \mathbf{H}_{IS}^{(ns)}$$

where $\mathbf{H}_{II}^{(0)}$ is the high field (ω_0) homonuclear (proton) dipolar Hamiltonian, $\mathbf{H}_{II}^{(ns)}$ is a nonsecular term (involves frequencies at $+2\omega_1$) and $\mathbf{H}_{IS}^{(ns)}$ is a nonsecular term which generally involves frequencies $+\omega_{1I} + \omega_{1S}$ (under the Hartman-Hahn conditions part of the IS interaction is restored to secularity). $\mathbf{H}_{SS}$ term (carbon-carbon dipolar interactions) is neglected as well as the all nonsecular terms corresponding to oscillations at the static field Larmor frequencies (ω_{0I}, ω_{0S}) [188].

When carbon (S) spins are locked following a CP 90° pulse, the $\mathbf{H}_{IS}$ term becomes nonsecular and, after these oscillations vanish, the S spin-lock magnetization is fractionally reduced. Thus, the observed decay is generally a sum of two Weibull functions, usually exponentials (Equation 20; $\mu = 1$): the initial slope reflects a $T_{1\varrho}$ dominated by a spin-lattice process and the final slope yields a $T_{1\varrho}$ dominated

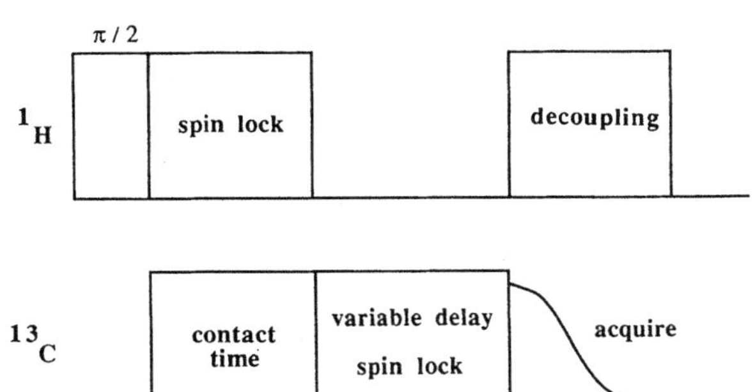

Fig. 43. ^{13}C $T_{1\varrho}$ pulse sequence with cross polarization

by spin-spin relaxation processes [4]. The contribution from the spin-spin relaxation arises from the modulation of the dipolar heteronuclear interactions by spontaneous rigid lattice proton spin fluctuations. This process occurs when strong $^{1}H-^{1}H$ static interactions are dominant, for example in crystalline polymers; it provides no information about molecular dynamics. Spin-lattice relaxation originates from the modulation of internuclear dipolar interactions by molecular motions at the rotating frame. The effect of proton static interactions in glassy polymers is very small as compared to the fluctuating dipolar fields associated with rotations of $^{13}C-^{1}H$ internuclear vectors. Thus, the T_1 mechanism for ^{13}C $T_{1\varrho}$ relaxation dominates in glassy polymers. In addition, the dynamic heterogeneity of solid polymers can also cause the observed $T_{1\varrho}$ for a given carbon to be a sum of different relaxation rates. Two component decays are observed in crosslinked poly(propylene oxide) while linear polymers show a simple exponential decay [201].

The fact that the relaxation in the rotating frame has contributions from both T_1 and T_2 mechanisms makes the interpretation of data on the molecular level rather complex. Furthermore, the determination of ^{13}C $T_{1\varrho}$ is complicated by the presence of a strongly coupled proton system (two components of experimental decay are not always clearly separated). The study of polymer motions in the kHz frequency range is of particular interest in the explanation of physical properties. An additional difficulty arises from ^{13}C $T_{1\varrho}$ sensitivity to the magic angle spinning frequencies which are of the same order order of magnitude as the $T_{1\varrho}$ motional frequencies (kHz range).

The ^{13}C $T_{1\varrho}$ study of cured epoxy resins at different radio frequency fields exhibit at rf frequencies above 40 kHz shorter values than those predicted for purely spin-spin effects. This finding indicates that the relaxation mechanism in the rotating frame at these rf fields is not determined exclusively by spin-spin mechanism, but reflect the spin-lattice effects. However, this conclusion is not generally valid for all polymers at room temperature. For example, ^{13}C $T_{1\varrho}$ in oriented polyethylene is found to be dominated by spin-spin effects even at a large rf field of 80 kHz. The temperature dependence of ^{13}C $T_{1\varrho}$ for different carbons in cured epoxy resin is pronounced only for the methyl group. The corresponding activation energy of 11 kJ/mol is also obtained from the proton relaxation studies in long alkanes [188].

The ^{13}C $T_{1\varrho}$ for crosslinked poly[oligo(ethylene glycol)dimethacrylate]s exhibits smaller values for all carbons in lightly crosslinked (rubbery gel) samples than the highly crosslinked (glassy) polymers. Only slightly smaller values observed for the glycol crosslink unit suggests that motion in the glycol crosslink is the same in all the samples [247]. The ^{13}C $T_{1\varrho}$ relaxation times for cured tetra(ethylene glycol)dimethacrylate PTEGDMA indicate the restriction of the relaxation mechanism after vitrification (Fig. 44). The lengthening of $T_{1\varrho}$ times after the vitrification is most pronounced for quaternary carbon (7-fold), less for the other carbon groups, while the relaxation of the CH_3 groups is much less restricted. Regardless of the relative contribution of spin-spin or spin-lattice relaxation to the the magnitude of $T_{1\varrho}$, the motional contribution (T_1) is rather significant, causing the damping of molecular motion in the relevant frequency range. It is concluded that the changes in relaxation rates arise from changes in the motion of the groups themselves, particularly as the group which shows the greatest change is unprotonated. Although there is no simple theory of molecular dynamics to interpret ^{13}C $T_{1\varrho}$ relaxation times, and unambigous

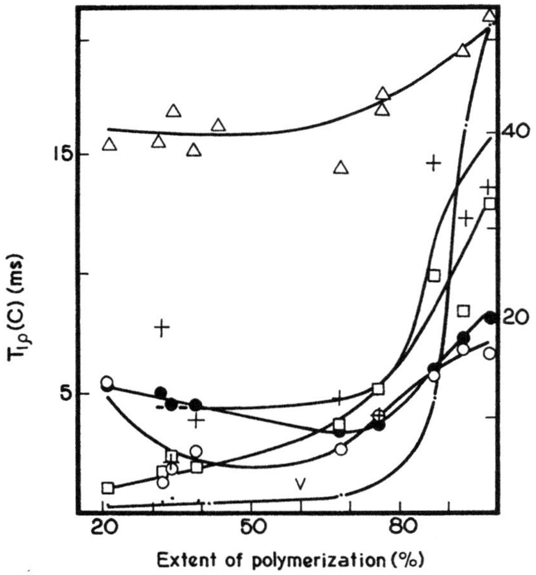

Fig. 44. Dependence of the ^{13}C $T_{1\varrho}$ relaxation times ($\gamma B_1 = 60$ kHz) for PTEGDMA on the extent of cure: inner CH_2O (●), end CH_2O (○), CH_2 (+) and CH_3 (△); for quaternary (□), represents $T_{1\varrho}/10$. Standard errors do not exceed the bounds of the symbols. The dots represent the compression modulus, E (MPa). V indicates the vitrification point (reprinted from ref. [248] with permission)

motional analysis is somewhat limited (small amplitude motions in tune could interfere with the large amplitude motions slightly out of tune) [4], the correlation of main chain relaxation times (in the kHz range) with some mechanical property, such as compression modulus, could be made [248].

All the carbons in the styrene crosslinked unsaturated polyesters based on fumaric acid, adipic acid and propylene glycol exhibit at room temperature the same qualitative dependence of ^{13}C $T_{1\varrho}$ on the curing temperature. Under the assumption that the relaxation times in the rotating frame are dominated by the T_1 relaxations (molecular motions) and that the larger $T_{1\varrho}$ values well below the glass transition indicate motional restrictions, the observed $T_{1\varrho}$ maximum for all the carbons of the network cured at 85 °C suggests that the most complete crosslinking reaction occurs at this particular temperature. A common pattern for all kinds of carbon atoms also indicates the mainly cooperative nature of the main-chain motion. The occurrence of the most complete reaction at 85 °C is in the agreement with the mechanical strength and the experimentally found relationship for reacted groups in the polyester. Above 85 °C the number of end groups increases with the cure temperature and, consequently, the molecular weight is decreasing. Thus, the observed maximum in $T_{1\varrho}$ is due to the combined effect of the highest crosslink density and highest carbon molecular weights [233].

The values of ^{13}C $T_{1\varrho}$ and their temperature dependence are found to be very sensitive to the nature of the build-up of the carbon magnetization: directly (by applying a 90° degree pulse), or indirectly by cross-polarization. The $^{13}C_{1\varrho}$ vs.

temperature dependence for a $CHCH_2$ resonance of a propylene oxide in the linear and crosslinked polyurethane networks (of different molecular mass) obtained by the CP pulse sequence (Fig. 43) and its GHPD analogue are shown in Fig. 45. The GHPD data for both, the linear and crosslinked polymer based on α,ω-dihydroxylpoly(propylene oxide) with M = 425, show no significant differences. However, a difference between CP and GHPD data for a linear polymer is observed, indicating a selective observation of the regions of different mobility. This difference is extremely pronounced in the crosslinked polymer. The $T_{1\varrho}$ for linear polymers shows, regardless of the applied technique, a minimum in the temperature region near or below 0 °C. The same minimum is present in the GHPD data for crosslinked polymer, while the two $T_{1\varrho}$ values obtained by the CP experiment on the network exhibit completely different behavior. The shorter $T_{1\varrho}$ in the network indicates the presence of some additional relaxation mechanisms such as a) large direct dipolar contributions to ^{13}C $T_{1\varrho}$ in the more rigid crosslinked region as compared to the more freely moving segments, b) contributions of phenyl motions of the crosslinking agent in the low temperature region and c) restricted rotations of the PPO methyl groups in the crosslinking region. With the increasing of M up to 3000, the difference between ^{13}C $T_{1\varrho}$ begins to diminish, indicating the more homogeneous structure of the later networks [201].

5.2.5 Dipolar-Dephasing (Interrupted Decoupling)

Since the ^{13}C—^1H dipolar interaction depends on an internuclear distance r_{CH} as $1/r^6_{CH}$, the coupling for proton bearing carbons (r ~ 1.1 Å) is much greater as

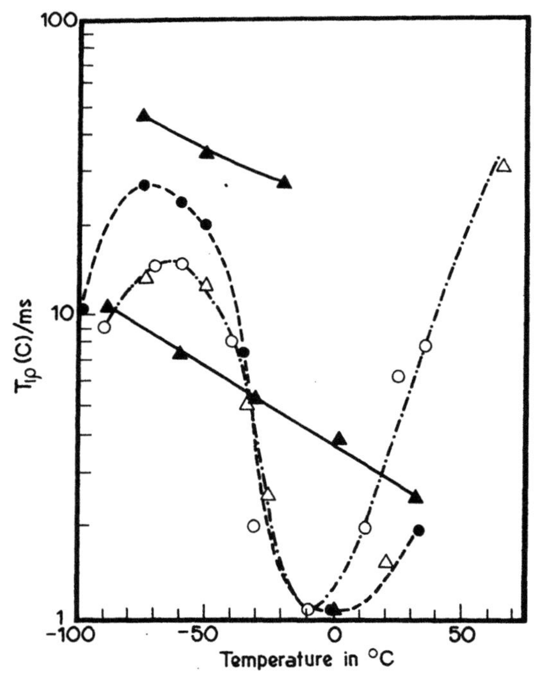

Fig. 45. Temperature dependence of ^{13}C $T_{1\varrho}$ for the propylene oxide $CHCH_2$ resonance in the linear (○, ●) and crosslinked polyurethanes (△, ▲). The values are determined from the pulse sequences with cross-polarization (●, ▲) and the direct application of a carbon 90° degree pulse (○, △) (adapted from Ref. [201])

compared to the protons on next neighboring carbon (r > 2.0 Å). The differentiation between carbon types is governed by the carbon dipolar relaxation rate (T_{1dd}):

$$\frac{1}{T_{1dd}} = \frac{N_H \gamma_H^2 \gamma_C^2 h \tau_c}{r_{CH}^6} \tag{52}$$

where N_H is the number of directly attached protons. Hence, it is possible to separate two different types of carbons by introducing a variable proton decoupling delay between the end of cross polarization process and the beginning of data acquisition [251]. The pulse sequence used in this experiment, called interrupted decoupling, dipolar dephasing or protonated carbon suppression, is shown in Fig. 46. During the delay period (T_{dd}) the carbon magnetization decays with the characteristic time constant T_2^* which depends on the strength of the heteronuclear dipolar interactions as well as on spin diffusion, molecular motion and MAS speeds. The decay of the signal intensity $I(T_{dd})$ generally follows Eq. (22) with $\mu_1 = 2$ and $\mu_2 = 1$ for short and long components, respectively. Thus, by changing T_{dd} values from short (about 10 μs) to long (approximately 400 μs) it should be possible to suppress the signals arising from carbons with different degrees of protonation.

The DD pulse sequence has been used for the study of cured furfuryl alcohol resins (Fig. 47). The assignment of four major peaks in cured resin (Fig. 47a) is the following: 2,5- and 3,4-positions of furfuryl repeating unit (151 and 109 ppm, respectively), CH_2 linkage of the repeating unit (29 ppm), and CH bridging group derived from CH_2 linkages (38 ppm). The DD spectrum with relatively long T_{dd} (100 μs) exhibits only one surviving peak at 151 ppm, due to the quaternary 2,5-carbons (Fig. 47b). Although the crosslinking mechanism is not exactly known, there are several suggestions for the structure of the cured resin. The appearance of only one peak in the DD spectrum indicates that there are no other quaternary carbons, which would be expected at about 120 ppm if the crosslinking occurs via 3,4-positions of the furan rings. On the basis of these results it can be concluded that a) crosslinking via the methylene bridges is the main crosslinking mechanism, and b) 3,4-substituted furan rings are not present in substantial quantities [234].

A differentiation between the two types of phenyl ring carbons (quaternary and those carbons which have protons directly bonded to them) in cured phenolic resins

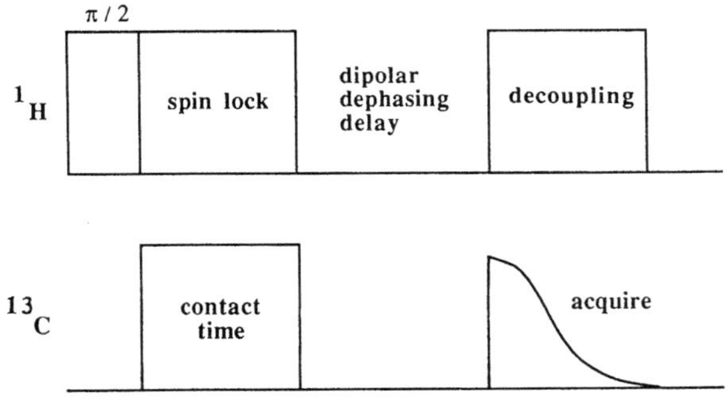

Fig. 46. Dipolar dephasing (DD) pulse sequence

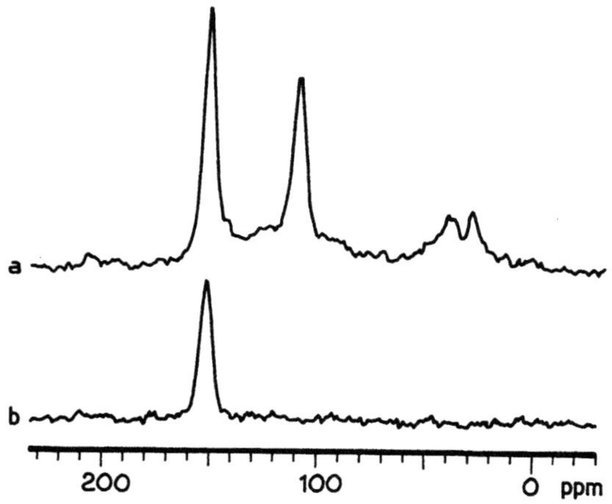

Fig. 47. 15 MHz ^{13}C MAS NMR spectrum of cured furfuryl alcohol resin **a**. The spectrum of the same resin obtained with dipolar dephasing (100 μs interrupted decoupling) **b** (reprinted from Ref. [234] with permission)

gives information on the number and positions of substitution on the phenyl ring introduced by crosslinks [216]. Furthermore, by using the DD pulse sequence a spectral deconvolution of overlapping aromatic signals in cured "high-*ortho*" and "random" phenolic resins [217], and assignments of absorptions in the spectra of cured polystyrylpyridine matrices [239] is also made. However, a clear separation of ^{15}N atoms in the urea-formaldehyde resin with directly and no directly attached protons, respectively, by using DD technique is not achieved [252]. On the contrary, a DD experiment for ^{15}N of a crosslinking agent, as well as for the ^{13}C atoms of a resin, is shown to be very useful for characterizing the intermediates in the curing reaction [218].

Quantitative analysis of dipolar dephasing spectra is not always possible because a) it is not easy to completely separate components: CH and CH_2 absorptions have very similar relaxation times and are suppressed at very short T_{dd}; furthermore, CH_3 groups, which have a greater amount of molecular mobility, relax slower and have similar relaxation times to quaternary carbons, b) the unsaturated carbon peak intensity could be lost to spinning sidebands and c) distortion of the spectra due to differential cross-polarization times for carbon atoms in different environments. However, these possibilities are found not to cause severe errors in the DD spectra of plasma polymerized hydrocarbons. The spectra are resolved into five broad components attributable to specific structural environments of crosslinked polymers and the mole fractions for each carbon type are determined with the reproducibility $\pm 15\%$ indicating the structural differences for the networks synthesized from a series of monomers (ethane, ethylene and acetylene). There was a decreasing hydrogen content from ethane to acetylene and approximately three times lower methyl group content in the acetylene polymerized networks [229, 230].

5.2.6 Distortionless Enhancement by Polarization Transfer

The distortionless enhancement by polarization transfer (DEPT) sequence [253] for the J-coupled heteronuclear spin system (J is $^{13}C—^{1}H$ coupling, approximately 135 to 170 Hz) is shown in Fig. 48. The pulse sequence is based on a resolvable spin-spin coupling between two nuclei, one of them (^{1}H) being the polarization source for the

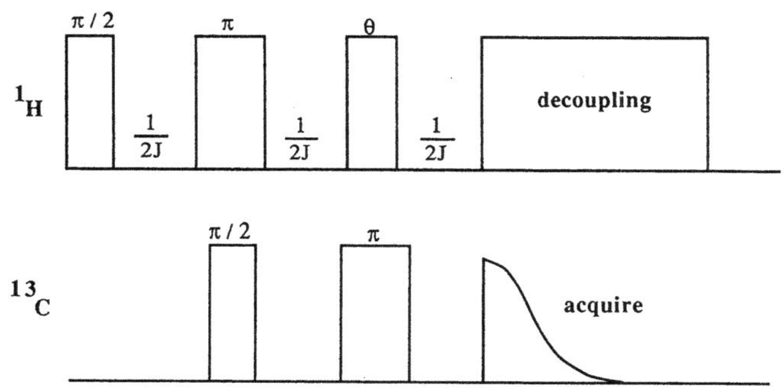

Fig. 48. Pulse sequence for the distortionless enhancement by polarization transfer (DEPT)

other (^{13}C). The different behavior of the particular multiplets (CH, CH_2 and CH_3) during the DEPT pulse sequence allows their separation into subspectra. The three DEPT CH_n (n = 1, 2, 3) subspectra are generated from the following combinations of experiments with the θ pulse settings 45°(θ_1), 90°(θ_2) and 135°(θ_3), respectively:

$$CH = \theta_2 - c(\theta_1 + a\theta_3)$$

$$CH_2 = \theta_1 - a\theta_3$$

$$CH_3 = \theta_1 + a\theta_3 - b\theta_2$$

Theoretical values of parameters a, b and c are 1.00, 0.707 and 0.00, respectively; however, experimental values are determined by obtaining the optimum cancellation of unwanted signals in the subspectra [254]. The DEPT pulse sequence has several advantages, such as insensitivity to J_{CH} and smaller sensitivity to rf-pulse inhomogeneities as compared to other analogous techniques. However, very accurate experimental conditions are generally required in order to obtain the final spectra, which are created from the three subspectra by a co-addition of the pure spectra. In the case of admixture of the subspectra, some additional linear combinations are usually required to produce the clean spectra.

The DEPT pulse sequence has been applied in the assignment of resonances which arise from the crosslinks in natural rubber. The aliphatic region of a ^{13}C NMR spectrum of vulcanized rubber and the three DEPT spectra, corresponding to the CH_3, CH_2 and CH groups, respectively, are shown in Fig. 49. The calculated chemical shifts for a number of possible network structural units indicate potential overlapping among the resonances arising from the different structures, as well as the presence of other types of crosslink structures. The ^{13}C chemical shift substituent effects due to a sulfur atom which is introduced at different positions in a polyisoprene repeating unit (and the corresponding modification of the repeating unit structure) predict, for example, a resonance from the various carbons in the region around 59 ppm. The CH DEPT subspectrum clearly shows that the absorption at 59 ppm has its origin in a CH structure, thus discriminating against the other possi-

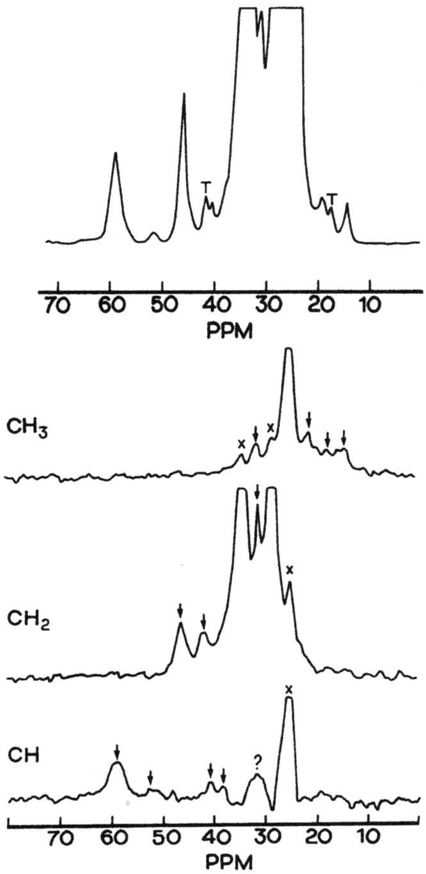

Fig. 49. Standard ^{13}C NMR spectrum (top) and DEPT spectra of sulfur cured natural rubber (6 h at 138 °C). The label "T" indicates peaks from trans-polyisoprene units, "X" marks residual peaks from other subspectra and arrows indicate peaks due to crosslink sites (adapted from Ref. [194])

bilities, and suggesting the presence of a sulfur atom directly bonded to a main chain via $>CH-$ tie point. Furthermore, the CH_2 nature of the absorption at 46 ppm suggests the existence of di- or polysulfidic crosslinks of the $-CH_2-S_n-$ type; at least three crosslink structures related to the different positions in a polyisoprene unit are involved. However, for a more precise determination of the proposed structures the identification of a number of its resonances is required, accompanied by a direct method for the detection of quaternary carbon atoms [194].

5.3 High Resolution Solid State NMR of Other Nuclei

5.3.1 ^{29}Si NMR

The polymer networks based on silicon are very suitable for the study of crosslinked systems because of the possibility to synthesize smaller model molecules and macromolecular networks of well defined structure. An additional advantage of silicon polymers is that the resonances arising for different structural units are usually well separated owing to the relatively large range of ^{29}Si NMR chemical shifts. Although some elastomeric networks could also be studied by solution NMR techniques, solid state NMR techniques are more useful because of the significantly shorter T_1 relaxation times.

As compared to the ^{13}C NMR of crosslinked poly(methyl siloxane) polymers, a direct observation of silicon nuclei by CP MAS in the particular samples yields more information about the network structure. Two ^{29}Si resonances in a sample prepared from CH_3SiCl_3 are ascribed to the general network repeating $CH_3Si^*(OSi\leqslant)_3$ unit and to the end-group structure $CH_3(OH)Si^*(OSi\leqslant)_2$, respectively; the two structural units, indicating a substantial quantity of end groups present, are not distinguishable in the corresponding ^{13}C NMR spectra. However, the ethoxy structures $CH_3CH_2OSi\leqslant$ are easier to detect in the carbon spectra than those of silicon. For this particular polymer, a combination of both techniques provides the most information [255].

CP spectra elastomeric PDMS networks recorded without MAS at low temperature (150 K) show powder patterns with signal areas from chains and crosslinks. Although the signal from chain silicons arises from an axially symmetric chemical shift tensor, its origin is not a consequence of rapid reorientation about a single axis, but rather from a coincidence in the two tensor principal values. The CP MAS spectra of two types of networks, vinyl- and hydroxyl-terminated systems, respectively, reveal their chemical structure, and the presence of irregularities in a real network. In a perfect network only one tetrafunctional silicon, $Si^*(O{-}Si{-}R_n)_4$, should be observed. However, both networks exhibit multiple resonances in this particular chemical shift region. In the hydroxyl-terminated network an additional signal is ascribed to the elastically ineffective inner-loop (the "bow-tie" structure) or to the trifunctional junctions. The vinyl-terminated networks show the same signal, and the additional one due to the double inner-loop ("butterfly" structure), a molecule which is not bound or entangled in the network. The ^{29}Si NMR chemical shifts for the chain silicons are found to be more sensitive to structural variations than the ^{13}C NMR shifts of bonded methylene groups [256, 257].

A ^{29}Si NMR has also proved to be a very powerful tool for investigations of the curing of polycarbosilanes, which consists of a skeleton of alternate carbon and silicon atoms. The NMR spectra of a linear polymer consist of two major resonances ascribed to the structural units Si^*C_4, HSi^*C_3, and very weak absorptions due to the units in which two silicon atoms are linked together: $Si^*Si_{(4-x)}C_x$, where $x = 1, 2,$ or 3. Crosslinking of the polymer chains by oxygen introduced by heating results in the appearence of two new absorptions ascribed to the Si*OSi and Si*OC crosslinking units; the relative signal intensity of the Si—O—Si unit increases linearly, while that of the SiC_3H unit decreases with the rise of curing temperature. It is concluded that oxygen atoms attack the SiC_3H bond, and convert it to the Si—O—Si bond [258].

5.3.2 ^{15}N NMR

A large number of polymer materials are based on nitrogen containing monomers, such as urea and related compounds, various amides, amines etc. The most frequent nitrogen forms found in polymers are primary, secondary and tertiary amides, respectively, the later usually being a part of the network structure. Since the nitrogen atoms are a part of different structural units: end-groups ($—NH_2$) main chain structure (—NH—) and network points ($-N\langle$), the direct observation of nitrogen nuclei is of particular interest. However, the low natural abundance of ^{15}N nuclei (only 0.37%) requires the use of ^{15}N-enriched compounds for NMR investigations.

Two ^{15}N-enriched urea-formaldehyde resins with different crosslink density were studied by the solid state CP MAS ^{15}N NMR. Despite at least six expected ^{15}N chemical shifts arising from tertiary, secondary and primary amides in the different structural moieties, both resins exhibit only two major peaks. The lower field resonance is more pronounced in the highly cured resin, suggesting its origin in the tertiary amides. A DD experiment, which would confirm this assumption, does not result in clearly separated secondary and tertiary amides. Thus, from the analytical point of view, it seems that ^{13}C NMR spectra are more useful than ^{15}N NMR spectra, although ^{15}N resonance data provide a useful supplement [252].

However, a complementary ^{15}N and ^{13}C NMR study of the hexamethylenetetramine (HMTA) curing of phenolic resins, using the isotope-enriched compounds was found to be very useful for the identification of intermediates and establishing the course of the curing process. The ^{13}C and the ^{15}N NMR spectra of the networks containing ^{13}C-enriched novolak phenolic resin and the ^{15}N-enriched HMTA, respectively, are shown in Fig. 50. The most dominant ^{13}C signal at lower temperatures (100 to 120 °C), assigned to HMTA (75 ppm), gradually decreases with increasing curing temperature, while the spectral intensity around 33 ppm significantly increases indicating that methylene crosslinks (para-para, para-ortho and ortho-ortho; from 40 to 30 ppm) formed during are crosslinking and originate from the ^{13}C-enriched HMTA. The new resonances appearing with increasing temperature at 83 and 56 ppm show a decrease in intensity at the later stages of reaction, indicating their origin in a reaction intermediate; the third one (167 ppm) remains essentially constant above 150 °C. The DD experiment reveals that they arise from carbon moieties with one or two directly attached hydrogens. The parallel ^{15}N NMR spectra show a broadening and decreasing of the main intensity around 40 ppm, which is, at the onset of cure, split into resonances at 44 and 41 ppm, respectively. Both of the resonances decrease during the cure, while the additional shoulder at 38 ppm increases (a few additional nitrogen resonances in the lower field region are also observed). The DD experiment confirms that ^{15}N resonances from 50 to 38 ppm are not due to the nitrogens with either one or two directly bonded hydrogens. The conclusions that i) the particular carbon resonances are due to the CH or CH_2 absorptions, and ii) those of the nitrogen do not have an origin in —NH— or $—NH_2$ groups, are compared with 15 prototype structures of the proposed intermediate and their respective ^{15}N and ^{13}C chemical shifts. A large number of the structures are eliminated from consideration because of the inadequate chemical shift range, or a different number of hydrogens attached to the nuclei under observation. In an additional comparison of the signal intensities during the curing, the new resonances are assigned to the benzoxazine and tribenzylamine type of structure, which originate from the transformations of the curing agent, HMTA. The fact that ^{13}C NMR resonances of tribenzylamine do not have the corresponding large ^{15}N NMR region of intensity around 52 ppm is explained in terms of the extreme sensitivity of ^{15}N nuclei to the geometrical environment, i.e. the structural heterogeneity and the resulting chemical shift dispersion [218].

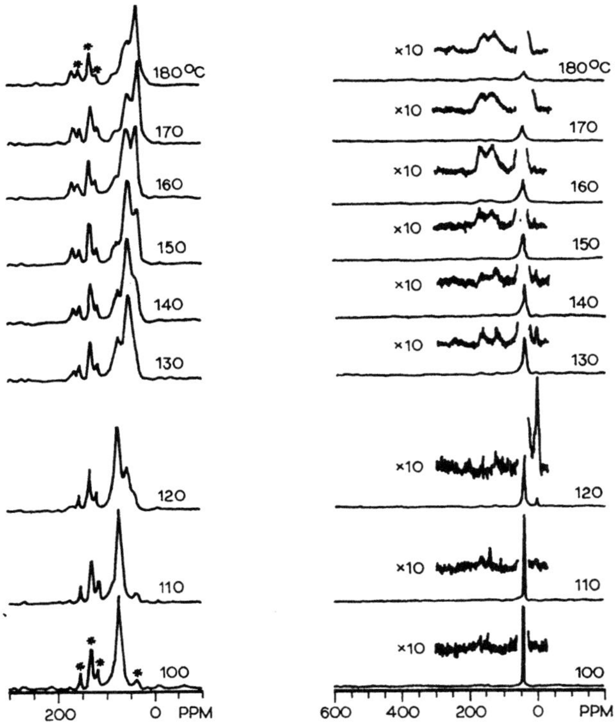

Fig. 50. CP MAS ^{13}C NMR spectra of phenolic resin cured with ^{13}C-labeled HMTA (left) and ^{15}N NMR spectra of the same resin cured with ^{15}N-labeled HMTA (right) for 1 h at the temperature indicated. Asterisks denote natural-abundance peaks (adapted from Ref. [218])

6 References

1. McCall, DW (1971) Acc. Chem. Res. 4: 223
2. McBrierty VJ (1974) Polymer 15: 503
3. Doskocilova D, Schneider B (1978) Adv. Coll. Interf. Sci. 9: 63
4. Schaefer J, Stejskal EO (1979) High-resolution ^{13}C NMR of solid polymers, in Levy GC (ed) Topics in Carbon-13 NMR spectroscopy, Wiley, New York, vol 3
5. McBrierty VJ, Douglass DC (1981) J. Polym. Sci., Macromol. Rev. 16: 295
6. McBrierty VJ (1983) Magn. Res. Rev. 8: 165
7. Havens JR, Koenig JL (1983) Appl. Spectrosc. 37: 226
8. Komoroski RA (ed) (1986) High resolution NMR spectroscopy of synthetic polymers in bulk. Deerfield Beach, Methods in stereochemical analysis, vol 7
9. Bauer DR (1986) Prog. Org. Coat 14: 45
10. Schaefer J, Stejskal EO (1976) J. Am. Chem. Soc. 98: 1031
11. Van Vleck JH (1948) Phys. Rev. 74: 1168
12. Gutowsky HS, Pake GE (1950) J. Chem. Phys. 18: 162
13. Holroyd LV, Mrowca BA, Guth E (1950) Phys. Rev. 79: 1026
14. Mrowca BA, Holroyd LV, Guth E (1950) Phys. Rev. 79: 1026
15. Holroyd LV, Codrington RS, Mrowca BA, Guth E (1951) J. Appl. Phys. 22: 696
16. Honnold VR, McCaffrey F, Mrowca BA (1954) J. Appl. Phys. 25: 1219

17. Burget J, Sącha J (1959) Czechosl. J. Phys. 9: 749
18. Gutowsky HS, Meyer LH (1953) J. Chem. Phys. 21: 2122
19. Gutowsky HS, Saika A, Takeda M, Woessner DE (1957) J. Chem. Phys. 27: 534
20. Chappell SE, Sauer JA, Woodward AE (1963) J. Polym. Sci., Part A, 1: 2805
21. Savel'eva ZI, Rakova SK, Kvlividze VI, Lezhnev NN (1972) Vysokomol. Soed. A14: 746
22. Wilson CW, Pake GE (1953) J. Polym. Sci. 10: 503
23. Slonim IY (1964) Vysokomol. Soed. 6: 1379
24. Shibayama K, Iisaka K (1966) Rep. Prog. Polym. Phys. Jpn. 9: 429
25. Iisaka K, Shibayama K (1976) Rep. Prog. Polym. Phys. Jpn. 19: 467
26. Slichter WP, Mandell ER (1958) J. Phys. Chem. 62: 334
27. Fuschillo N, Sauer JA (1957) J. Appl. Phys. 28: 1073
28. Fuschillo N, Sauer JA (1957) J. Chem. Phys. 26: 1348
29. Glick RE, Gupta RP, Sauer JA, Woodward AE (1960) Polymer 1: 340
30. Fujiwara S (1960) J. Polym. Sci. 44: 93
31. Tanaka K (1960) Bull. Chem. Soc. Jpn. 33: 1702
32. Jain PL (1959) J. Polym. Sci. 36: 443
33. Chuvaev VF, Ivanova LV, Zubov PI (1964) Vysokomol. Soed. A6: 1501
34. Karpov VL, Pomerantsev NM, Sergeev NM (1963) Vysokomol. Soed. A5: 100
35. Morita S, Shen M, Sawa G, Ueda M (1976) J. Polym. Sci., Polym. Phys. Ed. 14: 1917
36. Grad NM, Al'shits IM (1967) Vysokomol. Soed. A9: 832
37. Averko-Antonovich LA, Mukhutdinova TZ, Minkin VS, Yastrebov VN (1974) Vysokomol. Soed. A16: 1709
38. Huggins CM, St.Pierre LE, Bueche AM (1963) J. Polym. Sci., Part A 1: 2731
39. Barrante JR, Rochow EG (1964) J. Organomet. Chem., 1: 273
40. Model FS, Rochow EG (1970) J. Polym. Sci., Part A-2 8: 999
41. Tarasov VP, Smirnov YN, Yerofeyev LN, Irzhak VI, Rozenberg BA (1982) Vysokomol. Soed. A24: 2379
42. Banks L, Ellis B (1982) J. Polym. Sci., Polym. Phys. Ed. 20: 1055
43. Bamford CH, Eastmond GC, Whittle D (1975) Polymer 16: 377
44. Kelusky EC, Fyfe CA, McKinnon MS (1986) Macromolecules 19: 329
45. Dubault A, Deloche B, Herz J (1984) Polymer 25: 1405
46. Bloembergen N, Purcell EM, Pound RV (1948) Phys. Rev. 73: 679
47. Solomon I (1955) Phys. Rev. 99: 559
48. Kubo R, Tomita K (1954) J. Phys. Soc. Jpn. 9: 888
49. Jones GP (1966) Phys. Rev. 148: 332
50. Farrar TC, Becker ED (1971) Pulse and Fourier transform NMR, Academic, New York
51. Wardell GE, McBrietry VJ, Douglass DC (1974) J. Appl. Phys. 45: 3441
52. Derinovskii VS, Zakirov IN, Lantsov VM, Volkov VP, Rozenberg BA (1982) Vysokomol. Soed. A24: 2390
53. Larsen DW, Strange JH (1973) J. Polym. Sci., Polym. Phys. Ed. 11: 449
54. Rowland TJ, Labun LC (1978) Macromolecules 11: 466
55. Connor TM (1964) Trans. Faraday Soc. 60: 1574
56. Schaefer J (1973) Macromolecules 6: 882
57. Hunt BI, Powles JG (1966) Proc. Phys. Soc. 88: 513
58. Valeur B, Jarry JP, Geny F, Monnerie L (1975) J. Polym. Sci., Polym. Phys. Ed. 13: 667
59. Kimmich R (1977) Polymer 18: 233
60. De Gennes PG (1971) J. Chem. Phys. 55: 572
61. Howarth OW (1979) J. Chem. Soc., Faraday Trans. II, 75: 863
62. Woessner D (1962) J. Chem. Phys. 36: 1
63. Howarth OW (1978) J. Chem. Soc., Faraday Trans. II, 74: 1031
64. Jones AA, Stockmayer WH (1977) J. Polym. Sci., Polym. Phys. Ed. 15: 847
65. Bendler JT, Yaris R (1978) Macromolecules 11: 650
66. Skolnick J, Yaris R (1983) Macromolecules 16: 266
67. Eastmond GC, Smith EG (1977) Polymer 18: 244
68. Jurkiewicz A, Pislewski N, Kunert KA (1982) J. Macromol. Sci., Ser. Chem. A18: 511
69. Jurkiewicz A, Tritt-Goc J, Pislewski N, Kunert KA (1985) Polymer 26: 557

70. Nazarova II, Ol'khov YA, Nazarov VB, Baturin SM, Entelis SG (1975) Vysokomol. Soed., A17: 79
71. Nazarova II, Ol'khov YA, Baturin SM (1980) Vysokomol. Soed. A22: 391
72. Litvinov VM, Lavrukhin BD, Zhdanov AA, Andrianov KA (1977) Vysokomol. Soed. A19: 2031
73. Munie GC, Jonas J, Rowland TJ (1980) J. Polym. Sci., Polym. Chem. Ed. 18: 1061
74. Charlesby A, Folland R, Steven JH (1977) Proc. Roy. Soc. London, Ser. A, 355: 189
75. Gotlib, YY, Lifshits MI, Shevelev VA, Lishanskii IS, Balanina IV (1978) Vysokomol. Soed. A20: 423
76. Larsen DW, Strange JH (1973) J. Polym. Sci., Polym. Phys. Ed. 11: 65
77. Davis DD, Slichter WP (1973) Macromolecules 6: 728
78. Kunert KA, Chodak I, Ranachowski J, Pislewski N (1980) Polym. Bull. 3: 449
79. Kunert KA, Ranachowski J, Chodak I, Soszynska H, Pislewski N (1981) Polymer 22: 1677
80. Paci M, Del Vecchio E, Campana F (1986) Polym. Bull. 15: 21
81. Brown DR, Mnie GC, Jonas J (1982) J. Polym. Sci., Polym. Phys. Ed. 20: 1659
82. Marshall GL, Cudby MEA, Smith K, Stevenson TH, Packer KJ, Harris RK (1987) Polymer 28: 1093
83. Hahn EL (1950) Phys. Rev. 80: 580
84. Powles JG, Hartland A, Kail JAE (1961) J. Polym. Sci. 55: 361
85. Bergmann K, Demmler K (1974) Coll. Polym. Sci. 252: 193
86. Folland R, Charlesby A (1976) Int. J. Radiat. Phys. Chem. 8: 555
87. Folland R, Charlesby A (1977) Radiat. Phys. Chem. 10: 61
88. Charlesby A, Kafer P, Folland R (1978) Radiat. Phys. Chem. 11: 83
89. Folland R, Steven JH, Charlesby A (1978) J. Polym. Sci., Polym. Phys. Ed. 16: 1041
90. Charlesby A (1979) Radiat. Phys. Chem. 14: 919
91. Folland R, Charlesby A (1979) Polymer 20: 207
92. Folland R, Charlesby A (1979) Polymer 20: 211
93. Charlesby A, Folland R (1980) Radiat. Phys. Chem. 15: 393
94. Charlesby A, Jaroszkiewicz EM (1985) Eur. Polym. J. 21: 55
95. Bergmann K, Gerberding K (1981) Coll. Polym. Sci. 259: 990
96. Assink RA (1977) J. Polym. Sci., Polym. Phys. Ed. 15: 59
97. Goldman M, Shen L (1966) Phys. Rev. 144: 321
98. Assink RA, Wilkes GL (1981) J. Appl. Polym. Sci. 26: 3689
99. Barrett JL (1979) J. Radiat. Curing 6(4): 20
100. Andrianov KA, Zhdanov AA, Litvinov VM, Lavrukhin BD, Tsitsishvili VG (1975) Vysokomol. Soed. A17: 1323
101. Litvinov VM, Lavrukhin BD, Zhdanov AA, Andrianov KA (1976) Vysokomol. Soed. A18: 2515
102. Lind AC (1980) ACS, Div. Polym. Chem., Polym. Prepr. 21: 241
103. Nazarova II, Nazarov VB, Baturin SM (1982) Vysokomol. Soed. A24: 1724
104. Lantsov VM, Pakter MK, Abdrakhmanova LA, Zakirov IN, Paramonov YM, Zaitsev, YS (1986) Vysokomol. Soed. A28: 1047
105. Gotlib YY, Kuznetsova NN, Lifshits MI, Papukova KP, Shevelev VA (1974) Vysokomol. Soed. B16: 796
106. Kaufman S, Burger DJ (1970) J. Magn. Reson. 3: 218
107. Fujimoto K, Nishi T (1972) Nippon Gomu Kyokaishi 45: 828
108. Gotlib YY, Lifshits MI, Shevelev VA, Lishanskii IS, Balanina IV (1976) Vysokomol. Soed. A18: 2299
109. Stejskal EO, Tanner JE (1965) J. Chem. Phys. 42: 288
110. Meerwall E von, Palunas P (1987) J. Polym. Sci., Part B: Polym. Phys. 25: 1439
111. Doskocilova D, Schneider B (1982) Pure Appl. Chem. 54: 575
112. Doskovilova D, Schneider B, Trekoval J (1974) Coll. Czechoslov. Chem. Commun. 39: 2943
113. Doskocilova D, Schneider B, Jakes J (1978) J. Magn. Reson. 29: 79
114. Schneider B, Doskocilova D, Dybal J (1985) Polymer 26: 253
115. Flory PJ (1969) Statistical mechanics of chain molecules. Wiley, New York
116. Lifshits MI, Komarov EV (1983) Vysokomol. Soed A25: 2611
117. Lifshits MI (1987) Polymer 28: 454

118. Deloche B, Beltzung H, Hertz J (1982) J. Physique Lett. 43: L763
119. Cohen-Addad JP, Domard M, Herz J (1982) J. Chem. Phys. 76: 2744
120. Cohen-Addad JP, Dupeyre R (1983) Polymer 24: 400
121. Cohen-Addad JP, Viallat A (1986) Polymer 27: 1855
122. Dusek K, Spevacek J (1980) Polymer 21: 750
123. Spevacek J, Dusek K (1980) J. Polym. Sci., Polym. Phys. Ed. 18: 2027
124. Tsatskin ML, Bulai AK, Urman YG, Shipilevskii BA, Slonim IY (1975) Vysokomol. Soed. A17: 1748
125. Severnyi VV, Minas'yan, RM, Sokolov VV, Kisin AV (1976) Kauch. Rezina (2): 14
126. Nieto JL, Baselga J, Hernandez- Fuentes I, Llorente MA, Pierola IF (1987) Eur. Polym. J. 23: 551
127. Tibbitt JM, Shen M, Bell AT (1976) J. Macromol. Sci., Ser. Chem. A10: 1623
128. Gordon JE (1962) J. Phys. Chem. 66: 1150
129. De Villiers JP, Parrish JR (1964) J. Polym. Sci., Part A 2: 1331
130. Dinius RH, Choppin GR (1964) J. Phys. Chem. 68: 425
131. Creekmore RW, Reilley CN (1970) Anal. Chem. 42: 570
132. Darickova A, Doskocilova D, Sevcik S, Stamberg J (1970) J. Polym. Sci., Part B 8: 259
133. Tao DD, Doskocilova D, Stamberg J (1975) Angew. Makromol. Chem. 45: 139
134. Live D, Kent SBH (1982) ACS Symp. Ser. 193: 501
135. Schaefer J (1971) Macromolecules 4: 110
136. Ford WT, Yacoub SA (1981) J. Org. Chem. 46: 819
137. Errede LA, Newmark RA, Hill JR (1986) Macromolecules 19: 651
138. Marchenkov VV, Khitrin AK (1984) Khim. Fiz. 3: 1399
139. Yokota K, Abe A, Hosaka S, Sakai I, Saito H (1978) Macromolecules 11: 95
140. Ford WT, Balakrishnan T (1981) Macromolecules 14: 284
141. Mohanraj S, Ford WT (1985) Macromolecules 18: 351
142. Simon GP, Allen PEM, Williams DRG, Williams EH (1985) Eur. Polym. J. 21: 877
143. Bennett RL, Keller A, Stejny HH, Murray M (1976) J. Polym. Sci., Polym. Chem. Ed. 14: 3027
144. Bovey FA, Schilling FC, Cheng HN (1978) Adv. Chem. Ser. No. 169: 133
145. Randall JC, Zoepfl FJ, Silverman J (1983) Makromol. Chem., Rapid Commun. 4: 149
146. Marshall GL, Lander JA (1985) Eur. Polym. J. 21: 949
147. Marshall GL, Lander JA (1985) Eur. Polym. J. 21: 959
148. Marshall GL (1986) Eur. Polym. J. 22: 231
149. Birley AW, Dawkins JV, Kyriacos D, Bunn A (1981) Polymer 22: 812
150. Bergmark P, Flodin P (1987) Polymer 28: 1662
151. Paci M, Campana F (1985) Eur. Polym. J. 21: 717
152. Sayre JA, Assink RA, Lagasse RR (1981) Polymer 22: 87
153. Hodgkin JH, Willing RI, Eibl R (1981) J. Polym. Sci., Polym. Chem. Ed. 19: 1239
154. Taylor SK, Davidson F, Ovenall DW (1978) Photogr. Sci. Eng. 22: 134
155. Grenier-Loustalot M-F, Grenier P (1983) Makromol. Chem. 184: 2477
156. Grenier-Loustalot M-F, Grenier P (1984) J. Polym. Sci., Polym. Chem. Ed. 22: 4011
157. Grenier-Loustalot M-F, Cazaux F, Grenier P (1984) Eur. Polym. J. 20: 1137
158. Grenier-Loustalot M-F, Orozoco L, Grenier P (1986) Eur. Polym. J. 22: 921
159. Pastoriza MA, Bertorello HE (1986) Polym. Bull. 15: 241
160. Nikolaev BP, Pasechnik VA, Petrov LN, Polyakova YA (1983) Vysokomol. Soed. A25: 2285
161. Asakura T, Suzuki K, Horie K (1981) Makromol. Chem. 182: 2289
162. Doskocilova D, Schneider B, Jakes J (1980) Polymer 21: 1185
163. Jones AJ, Leznoff CC, Svirskaya PI (1982) Org. Magn. Res. 18: 236
164. Ohtani N, Wilkie CA, Nigam A, Regen SL (1981) Macromolecules 14: 516
165. Ford WT (1984) J. Polym. Sci., Polym. Chem. Ed. 22: 509
166. Sternlicht H, Kenyon GL, Packer EL, Sinclair J (1971) J. Am. Chem. Soc. 93: 199
167. Beshah K, Mark JE, Ackerman JL, Himstedt A (1983) Proc. Am. Chem. Soc., Div. Polym. Mater. Sci. Eng. 49: 288
168. Evans TL, Patterson DB, Suszko PR, Allcock HR (1981) Macromolecules 14: 218
169. Andrew ER, Bradbury A, Eades RG (1959) Nature (London) 183: 1802
170. Lowe IJ (1959) Phys. Rev. Lett. 2: 285

171. Andrew ER, Jasinski A (1971) J. Phys. C., Solid State Phys. 4: 391
172. Schneider B, Doskocilova D, Babka J, Ruzicka Z (1980) J. Magn. Reson. 37: 41
173. Doskocilova D, Schneider B (1972) Macromolecules 5: 125
174. Tao DD, Doskocilova D, Stamberg J (1974) Angew. Makromol. Chem. 38: 129
175. Carr HY, Purcell EM (1954) Phys. Rev. 94: 630
176. Gerstein BC, Dybowski CR (1985) Transient techniques in NMR of solids, An introduction to theory and practice, Academic, Orlando
177. Waugh JS, Huber LM, Haeberlen U (1968) Phys. Rev. Lett. 20: 180
178. Mansfield P, Ware D (1968) Phys. Rev. 168: 318
179. Waugh JS, Wang CH (1967) Phys. Rev. 162: 209
180. Geschke D, Poschel K, Doskocilova D, Schneider B (1985) Acta. Polym. 36: 645
181. Geschke D, Poschel K (1986) Coll. Polym. Sci. 264: 482
182. Mansfield P (1971) J. Phys. C: Solid State Phys. 4: 1444
183. Rhim W-K, Ellemann DD, Vaughan RW (1973) J. Chem. Phys. 59: 3740
184. Dybowski CR, Vaughan RW (1975) Macromolecules 8: 50
185. Dybowski CR, Vaughan RW (1977) in Labana SS (ed) Chemistry and properties of crosslinked polymers, Academic, New York, pp. 547–558
186. Gerstein BC (1981) Phil. Trans. Roy. Soc. London, Ser. A 299: 521
187. Okada T (1975) Rep. Prog. Polym. Phys. Jpn. 18: 479
188. Garroway AN, Moniz WB, Resing HA (1979) ACS Symp. Ser., No. 103: 67
189. Schaefer J, Chin SH, Weissman SI (1972) Macromolecules 5: 798
190. Komoroski RA, Mandelkern L (1976) J. Polym. Sci., Polym. Lett. Ed. 14: 253
191. Kentgens APM, Veeman WS, van Bree J (1987) Macromolecules 20: 1234
192. Terao T, Miura H, Saika A (1981) J. Chem. Phys. 75: 1573
193. Patterson DJ, Koenig JL, Shelton JR (1983) Rubber Chem. Technol. 56: 971
194. Komoroski RA, Shockcor JP, Gregg EC, Savoca JL (1986) Rubber Chem. Technol. 59: 328
195. Koenig JL, Patterson DJ (1986) Elastomerics 118(11): 21
196. Zaper AM, Koenig JL (1987) Rubber Chem. Technol. 60: 252
197. Zaper AM, Koenig JL (1987) Rubber Chem. Technol. 60: 278
198. Patterson DJ, Koenig JL (1987) Appl. Spectrosc. 41: 441
199. Patterson DJ, Koenig JL (1984) ACS Symp. Ser. No. 243: 205
200. Andreis M, Liu J, Koenig JL (submitted to J. Polym. Sci., Polym. Phys. Edn. for publ.) Solid state carbon-13 NMR studies of vulcanized elastomers. V. Observation of new structures in sulfur vulcanized natural rubber
201. Dickinson LC, Morganelli PL, MacKnight WJ, Chien JCW (1987) Makromol. Chem., Rapid Commun. 8: 425
202. Bauer DR, Dickie RA, Koenig JL (1984) J. Polym. Sci., Polym. Phys. Ed. 22: 2009
203. Bauer DR, Dickie RA, Koenig JL (1986) Ind. Eng. Chem., Prod. Res. Dev. 25: 289
204. Komoroski RA (1983) J. Polym. Sci., Polym. Phys. Ed. 21: 2551
205. Vanderhart DL, Earl WL, Garroway AN (1981) J. Magn. Reson 44: 361
206. Periyasamy M, Ford WT (1987) Polym. Mater. Sci. Eng. 56: 184
207. Andreis M, Koenig JL (submitted to Rubber Chem. Technol. for publ.) Solid state carbon-13 NMR studies of vulcanized elastomers. VI. Relaxation in sulfur vulcanized natural rubber
208. Hartman SR, Hahn EL (1952) Phys. Rev. 128: 2042
209. Pines A, Gibby MG, Waugh JS (1973) J. Chem. Phys. 59: 569
210. Earl WL, Vanderhart DL (1979) Macromolecules 12: 762
211. Maciel GE, Szeverenyi M, Early TA, Myers GE (1983) Macromolecules 16: 598
212. Wong AC, Garroway AN, Ritchey WM (1981) Macromolecules 14: 832
213. Dixon WT (1981) J. Magn. Reson 44: 220
214. Sefcik MD, Stejskal EO, McKay RA, Schaefer J (1979) Macromolecules 12: 423
215. Patterson DJ, Shields CM, Cholli A, Koenig JL (1984) Polym. Prepr. 25: 358
216. Fyfe CA, Rudin A, Tchir W (1980) Macromolecules 13: 1320
217. Bryson RL, Hatfield GR, Early TA, Palmer AR, Maciel GE (1983) Macromolecules 16: 1669
218. Hatfield GR, Maciel GE (1987) Macromolecules 20: 608
219. Balimann GE, Groombridge CJ, Harris RK, Packer KJ, Say BJ, Tanner SF (1981) Phil. Trans. Roy. Soc. Lond. A299: 643
220. Garroway AN, Ritchey WM, Moniz WB (1982) Macromolecules 15: 1051

221. Cholli A, Ritchey WM, Koenig JL (1984) ACS Symp. Ser. No. 243: 234
222. Grenier-Loustalot M-F, Grenier P (1986) Eur. Polym. J. 22: 457
223. Grenier-Loustalot M-F, Cazaux F, Berecoechea J, Grenier P (1986) Eur. Polym. J. 22: 471
224. Chalmers AA, Perlstein P (1987) Eur. Polym. J. 23: 887
225. Sohma J, Shiotani M, Yoshida T (1981) Rep. Prog. Polym. Phys. Jpn. 24: 497
226. Shiotani M, Murakami S, Sohma J, Yoshida T (1982) Rep. Prog. Polym. Phys. Jpn. 25: 593
227. Sohma J, Shiotani M, Murakami S (1983) Radiat. Phys. Chem. 21: 413
228. O'Donnell JH, Whittaker AK (1985) Br. Polym. J. 17: 51
229. Kaplan S, Dilks A (1981) Thin Solid Films 84: 419
230. Dilks A, Kaplan S, Van Laeken A (1981) J. Polym. Sci., Polym. Chem. Ed. 19: 2987
231. Kaplan S, Dilks A (1983) J. Polym. Sci., Polym. Chem. Ed. 21: 1819
232. Paci M, Crescenzi V, Campana F (1982) Polym. Bull. 7: 59
233. Bergmark P, Flodin P (1987) Polymer 28: 1657
234. Maciel GE, Chuang I-S, Myers GE (1982) Macromolecules 15: 1218
235. Chuang I-S, Maciel GE, Myers GE (1984) Macromolecules 17: 1087
236a. English AD, Chase DB, Spinelli HJ (1983) Macromolecules 16: 1422; b. English AD, Spinelli HJ (1984) ACS Symp. Ser. 243: 257
237. Bauer DR, Dickie RA, Koenig JL (1985) Ind. Eng. Chem., Prod. Res. Dev. 24: 121
238. Barron PF, O'Donnell JH, Whittaker AK (1985) Polym. Bull. 14: 339
239. Laupretre F, Monnerie L, Bloch B (1986) Anal. Chim. Acta 189: 117
240. Fyfe CA, McKinnon MS, Rudin A, Tchir WJ (1983) J. Polym. Sci., Polym. Lett. Ed. 21: 249
241. Fyfe CA, McKinnon MS, Rudin A, Tchir WJ (1983) Macromolecules 16: 1216
242. Maciel GE, Chuang I-S, Gollob L (1984) Macromolecules 17: 1081
243. Fyfe CA, Bemi L, Childs R, Clark HC, Curtin D, Davies J, Drexler D, Dudley RL, Gobbi GC, Hartman JS, Hayes P, Klinowski J, Lenkinski RE, Lock CJL, Paul IC, Rudin A, Tchir W, Thomas JM, Wasylishen RE (1982) Phil. Trans. Roy. Soc. London, Ser. A 305: 591
244. So S, Rudin A (1985) J. Polym. Sci., Polym. Lett. Ed. 23: 403
245. Earnshaw RG, Price CA, O'Donnell JH, Whittaker AK (1986) J. Appl. Polym. Sci. 32: 5337
246. Havens JR, Thakur M, Lando JB, Koenig JL (1984) Macromolecules 17: 1071
247. Allen PEM, Simon GP, Williams DRG, Williams EH (1984) Polym. Bull. (Berlin) 11: 593
248. Allen PEM, Simon GP, Williams DRG, Williams EH (1986) Eur. Polym. J. 22: 549
249. Haw JF, Johnson NA (1986) Anal. Chem. 58: 3254
250. Curran SA, Padwa AR (1987) Macromolecules 20: 625
251. Opella SJ, Frey MH (1979) J. Am. Chem. Soc. 101: 5854
252. Chuang I-S, Hawkins B, Maciel GE, Myers GE (1985) Macromolecules 18: 1482
253. Doddrell DM, Pegg DT, Bendall MR (1982) J. Magn. Res. 48: 323
254. Pegg DL, Doddrell DM, Bendall MR (1979) J. Chem. Phys. 63: 309
255. Maciel GE, Sullivan MJ, Sindorf DW (1981) Macromolecules 14: 1607
256. Beshah K, Mark JE, Ackerman JL, Himstedt A (1986) J. Polym. Sci., Part B, Polym. Phys. 24: 1207
257. Beshah K, Mark JE, Ackerman JL (1986) Macromolecules 19: 2194
258. Taki T, Maeda S, Okamura K, Sato M, Matsuzawa T (1987) J. Mater. Sci. Lett. 6: 826

Editor: H.-J. Cantow
Received Juli 21, 1988

Author Index Volumes 1–89

Subject Index

GPSR Compliance

The European Union's (EU) General Product Safety Regulation (GPSR) is a set of rules that requires consumer products to be safe and our obligations to ensure this.

If you have any concerns about our products, you can contact us on ProductSafety@springernature.com

In case Publisher is established outside the EU, the EU authorized representative is:

Springer Nature Customer Service Center GmbH
Europaplatz 3
69115 Heidelberg, Germany

Batch number: 10620881

Printed by Printforce, the Netherlands